WATER & WASTEWATER TREATMENT:
A Guide for the Nonengineering Professional

Lake Redman Reservoir, York County, PA.

WATER & WASTEWATER TREATMENT: A Guide for the Nonengineering Professional

Joanne E. Drinan

Technical Writer, Wastewater Treatment Plant Operations

ASSOCIATE EDITOR

Nancy E. Whiting

Technical Writer, ENCORE Technical Resources, Inc.

TECHNOMIC
PUBLISHING CO., INC.
LANCASTER • BASEL

Water and Wastewater Treatment
aTECHNOMIC®publication

Technomic Publishing Company, Inc.
851 New Holland Avenue, Box 3535
Lancaster, Pennsylvania 17604 U.S.A.

Printed in the United States of America
10   9   8   7   6   5   4   3   2   1

Main entry under title:
    Water and Wastewater Treatment: A Guide for the Nonengineering Professional

A Technomic Publishing Company book
Bibliography: p.
Includes index p. 311

Library of Congress Catalog Card No. 00-109963
ISBN No. 1-58716-049-8

HOW TO ORDER THIS BOOK

BY PHONE: 800-233-9936 or 717-291-5609, 8AM–5PM Eastern Time

BY FAX: 717-295-4538

BY MAIL: Order Department
Technomic Publishing Company, Inc.
851 New Holland Avenue, Box 3535
Lancaster, PA 17604, U.S.A.

BY CREDIT CARD: American Express, VISA, MasterCard

BY WWW SITE: http://www.techpub.com

To Mike Drinan and to John and Martha Goeke

Table of Contents

Preface

THIS book, *Water and Wastewater Treatment: A Guide for the Nonengineering Professional*, presents all the basic unit processes involved in drinking water and wastewater treatment, step-by-step, in jargon-free language. It describes each unit process, what function the process provides in water or wastewater treatment, and the basic equipment each process uses. It details how the processes fit together within a drinking water or wastewater treatment system, and surveys the fundamental concepts that make up water/wastewater treatment processes as a whole.

Designed to cover the specific needs of nonengineers in the water and wastewater industry, this book is a useful training tool for those who need to be knowledgeable about water and wastewater processes, techniques, and equipment, but do not work directly with day-to-day treatment plant operations. Municipal managers, departmental and administrative assistants, equipment sales or marketing personnel, and customer service representatives, as well as those on utility municipality boards and those new to the water and wastewater field will find *Water & Wastewater Treatment* a useful resource.

By design, this text does not include mathematics, engineering, chemistry, or biology. However, it does include numerous illustrations and photos, as well as an extensive glossary of terms and abbreviations for easy comprehension of concepts and processes, and for quick reference.

Water and Wastewater Treatment is formatted in four parts: Part I covers the basics of the hydrogeologic cycle; Part II covers the basics of water treatment; Part III covers the basics of wastewater treatment; and Part IV covers water and wastewater biosolids management and disposal, providing the reader with a simple and direct guidebook from start to finish in water and wastewater treatment.

xiii

Falling Spring, VA.

Acknowledgements

WE offer special thanks to Dalvin Crug at the Susquehanna Plant of the Lancaster, PA Bureau of Water, Gene Hecker at the Conestoga River Water Treatment Plant, and Bill Horst at the Lancaster, PA Advanced Wastewater Treatment Plant, for allowing us to photograph the Plants' facilities, as well as to Mike Parcher for his generosity and advice. Special thanks are also due to Frank R. Spellman for his writings and instruction that made this work possible, and for the many photographs he willingly loaned.

Introduction: Problems Facing Water and Wastewater Treatment Management

MANAGEMENT FOR WATER AND WASTEWATER TREATMENT FACILITIES

W ATER and Wastewater treatment facilities are usually owned, operated, and managed by the community (the municipality) where they are located. While many of these facilities are privately owned, the majority of Water Treatment Plants (WTPs) and Wastewater Treatment Plants (WWTPs) are Publicly Owned Treatment Works (POTWs).

These publicly owned facilities are managed on site by professionals in the field. On-site management, however, is usually controlled by a board of elected, appointed, or hired directors/commissioners, who set policy, determine budget, plan for expansion or upgrading, hold decision-making power for large purchases, and in general, control the overall direction of the operation.

When final decisions on matters that affect plant performance are in the hands of, for example, a board of directors comprised of elected and appointed city officials, their knowledge of the science, engineering, and hands-on problems that those who are on site must solve can range from "all" to "none." Matters that are of critical importance to those in on-site management may mean little to those on the Board. The Board of Directors may indeed also be responsible for other city services, and have an agenda that encompasses more than just the water or wastewater facility. Thus, decisions that affect on-site management can be affected by political and financial concerns that have little to do with the successful operation of a POTW.

Finances and funding are always of concern, no matter how small or large, well-supported or under-funded the municipality. Publicly owned treatment works are generally funded from a combination of sources. These include local taxes, state and federal monies (including grants and matching funds for upgrades), as well as usage fees paid by water and wastewater customers. In smaller communities, the W/WW plants may be the only city services that actually generate income. This is especially true in water treatment and delivery, which is commonly looked upon as the "cash cow" of city services. Funds generated by the facility do not always stay with the facility. These funds can be re-assigned to support other city services—and when facility upgrade time comes, funding for renovations can be problematic.

MANAGEMENT PROBLEMS FACING POTWS

Problems come and go, shifting from year to year and site to site. They range from the problems caused by natural forces (drought, storms, earthquake, fire, and flood) to those caused by social forces.

In general, four areas are of constant concern in many facilities:

- complying with regulations, and coping with new and changing regulations
- maintaining and upgrading facilities, unit processes and equipment
- staving off privatization
- maintaining a viable and well-trained workforce

COMPLIANCE WITH NEW, CHANGING, AND EXISTING REGULATIONS

Adapting the facility and workforce to the challenges of meeting changing regulations is of concern in both water and wastewater treatment. The Clean Water Act Amendments that went into effect in February of 2001 require water treatment plants to meet tougher standards, presenting new problems for treatment facilities to deal with, and offering some possible solutions to the problems of meeting the new standards. These regulations provide for communities to upgrade their treatment systems, replacing aging and outdated equipment with new process systems. Their purpose is to ensure that facilities are able to filter out higher levels of impurities from drinking water, thus reducing the health risk from bacteria, protozoa, and viruses, and are able to decrease levels of turbidity, and reduce concentrations of chlorine by-products in drinking water.

The National Pollution Discharge Elimination System program (NPDES) established by the Clean Water Act, issues permits that control wastewater treatment plant discharges. Meeting permit is always of concern for wastewater treatment facilities, because the effluent discharged into water bodies affects those downstream of the release point. Individual point source dischargers must use the best available technology (BAT) to control the levels of pollution in the effluent they discharge into streams. As systems age, and best available technology changes, meeting permit with existing facilities becomes increasingly difficult.

UPGRADING FACILITIES, EQUIPMENT AND UNIT PROCESSES

Many communities are facing problems caused by aging equipment, facilities, and infrastructure. The inevitable decay caused by system age is compounded by increasing pressure on inadequate older systems to meet demands of increased population and urban growth. Facilities built in the 1960s and 1970s are now 30 to 40 years old, and not only are they showing signs of wear and tear, they simply were never planned to handle the level of growth that has occurred in many municipalities.

Regulations often provide a reason to upgrade. By matching funds or providing federal monies to cover some of the costs, municipalities can take advantage of a window of opportunity to improve their facility at a lower direct cost to the community. Those federal dollars, of course, do come with strings attached; they are meant to be spent on specific projects in specific areas.

Changes in regulation may force the issue. The use of chlorine as a disinfectant is under close scrutiny now, and pressure to shift to other forms of disinfectant is increasing. This would mean replacing or changing existing equipment, increased chemical costs, and could easily involve increased energy and personnel costs. Equipment condition, new technology, and financial concerns are all up for consideration when upgrades or new processes are chosen. Also of consideration is the safety of the process, because of the demands made by OSHA and the EPA in their Process Safety Management/Risk Management Planning regulations. The potential of harm to workers, the community, and the environment are all under study, as are the possible long-term effects of chlorination on the human population.

STAVING OFF PRIVATIZATION

Privatization is becoming of greater and greater concern. Governance Boards see privatization as a potential way to shift liability and responsibility from the municipality's shoulders, with the attractive bonus of cutting costs. Both water and wastewater facilities face constant pressure to work more efficiently, more cost-effectively, with fewer workers, to produce a higher quality product; that is, all functions must be value-added. Privatization is increasing, and many municipalities are seriously considering outsourcing parts or all of their operations to contractors.

On-site managers often consider privatization threatening. In the worse case scenario, a private contractor could bid the entire staff out of their jobs. In the best case? Privatization is often a very real threat that forces on-site managers into workforce cuts, improving efficiency and cutting costs—while they work to ensure the community receives safe drinking water and the facility meets standards and permits, with fewer workers—and without injury or accident to workers, the facility, or the environment. While cutting out dead wood in an organization is sometimes needed, concerns over privatization can cause cuts that go too deep for safe work practices.

Valid concerns about how privatization can affect the municipality's and environment's safety are common, with the argument that outsiders will not have the same level of concern for the community as local management. But the bottom line is simple. No matter what the costs, the community's need for safe water supplies and effective waste treatment and disposal is paramount.

MAINTAINING A VIABLE WORKFORCE

Maintaining a viable, well-trained workforce becomes ever more difficult as new regulations require higher levels of training and certification for workers. Low unemployment rates also increase an employee's opportunity to move from job to job, seeking higher pay. Municipalities are often tied to state or city worker payment levels, and can offer little flexibility for pay increases. Workers who received solid training financed by the municipality can sometimes simply take their certification and walk into a higher-paid position elsewhere, because of standard contract or employment policy limitations and an inflexible pay structure. When new regulations mandating worker certification for water treatment are eminent, an already trained, skillful, knowledgeable worker is an attractive target.

SUMMARY

Many problems face those who operate treatment facilities, but their most critical concern is a basic one: providing the best level of treatment possible, ensuring the safe condition of our water supplies.

PART I

THE HYDROGEOLOGIC CYCLE

Tucqvan glen, Pennsylvania.

CHAPTER 1

Natural and Urban Water Cycles

1.1 HYDROGEOLOGIC PATHWAYS

W ATER travels constantly throughout our world, evaporating from the sea, lakes, and soil and transpiring from foliage. Once in the atmosphere, water travels for miles as vaporous clouds, then falls again to earth as rain or snow. On solid ground again, water runs across the land or percolates into the soil to flow along rock strata into aquifers. As groundwater, its movement continues, slower perhaps, but certainly not static. Whether trickling through the matrix of rock and soil underground, drifting across the sky as clouds, or rolling across the land as a river, water is in constant motion. Eventually water finds its way again to the sea, and the cycle continues.

1.2 THE WATER CYCLE

In the natural water cycle or *hydrological cycle,* water in its three forms— solid, liquid, and vapor—circulates through the biosphere. The hydrologic cycle describes water's circulation through the environment. Evaporation, transpiration, runoff, and precipitation (terms which describe specific water movements) are the means by which water travels from bodies of water or soil to the atmosphere and back again—from surface water to water vapor, perhaps to ice, and back again to surface water (see Figure 1.1).

Water leaves the earth's surface and enters the atmosphere either by evaporation from bodies of water on the earth's surface (lakes, rivers, and oceans) or through the transpiration of plants. In the atmosphere, water vapor forms into clouds. As the air pressure and temperature change, the vapor in the clouds

7

Johnson Shut-Ins. The Black River, Missouri.

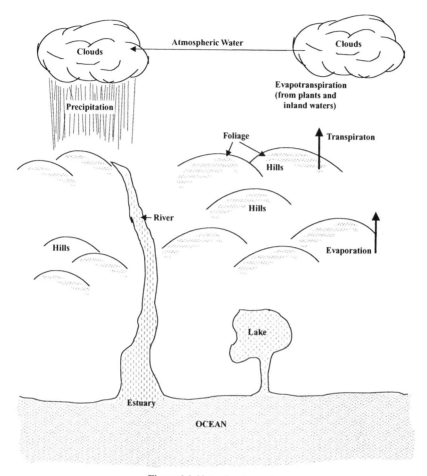

Figure 1.1 Natural water cycle.

condenses, depositing moisture in the form of precipitation on the land and sea. If we were able to follow a single drop of water on its journey, the trail of that drop of water could cover thousands of miles between evaporation and precipitation. Our drop of water would fall as precipitation onto the face of the earth, and would flow in streams and rivers to the ocean, or seep into the earth to become groundwater. Even groundwater eventually flows toward the ocean for recycling.

By this process, water is constantly cycled and recycled, cleaned and re-purified by the long journey into the atmosphere, back to earth, through river courses, and into the soil. Coupled with dilution and the natural processes of self-purification, the hydrogeologic system works well to maintain good stream

TABLE 1.1. World Water Distribution.

Location	Percent of Total
Land areas	
Freshwater lakes	0.009
Saline lakes and inland seas	0.008
Rivers (average instantaneous volume)	0.0001
Soil moisture	0.005
Groundwater (above depth of 4000 m)	0.61
Ice caps and glaciers	<u>2.14</u>
	2.8
Atmosphere (water vapor)	0.001
Oceans	<u>97.3</u>
Total all locations (rounded)	**100.0**

Source: Adapted from Peavy et al., *Environmental Engineering.* New York: McGraw-Hill, 1985, p. 12.

quality, and if polluted, to restore water and stream flow to its normal, healthy state.

We use and re-use the same water we have always used; our water supply, no more or less water than the earth has always held, is finite, a constantly recycled resource. Of all the water within our reach, only a small fraction is not salt, and much of the fresh water is not available to us (see Table 1.1). However, we are also using more of the small percentage of pure water available to us than ever before, and the water we use—and the water we have used—is not always safe to return to the environment. To maintain our water supplies, we must work to ensure that our wastes do not overwhelm the natural self-purification process the hydrogeologic cycle provides. We thus treat the water we use, compressing the self-purification ability demonstrated by nature into an artificial water cycle— the urban water cycle.

1.2.1   STREAM SELF-PURIFICATION

Healthy watercourses possess a balance of biological organisms that aid in disposing of small amounts of pollution. This balance of organisms means that streams with a normal pollution load "self-purify" within definable stream zones past the point of pollution (see Figure 1.2).

Upstream of the point of pollution entry is the "clean zone." At the pollution discharge point, the stream becomes turbid in the " zone of recent pollution." The pollution in the stream causes dissolved oxygen (DO) levels to sharply drop in the "septic zone." As organic wastes decompose, the stream quality begins to return to normal levels in the "recovery zone" (see Table 1.2). With enough time and no further pollutants added to the stream, the stream will return to

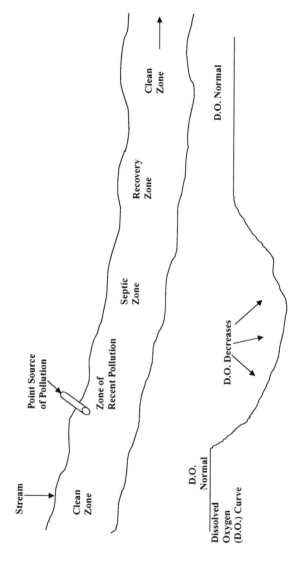

Figure 1.2 Stream zones (above) and dissolved oxygen curve (below).

11

TABLE 1.2. Stream Zone Conditions.

Clean Zone	Zone of Recent Pollution	Septic Zone	Recovery Zone
High DO	DO varies with organic load	Little or no DO	DO from 2 ppm to saturation
Low BOD	High BOD	High, decreasing BOD	Lower BOD
No turbidity	High turbidity	High turbidity	Less turbidity
Water surface normal, clear water	Water dark	Water surface appears oily, water dark	Water surface normal, light green tinge
No gas bubbles	Forming gases	Rising gas bubbles	No gas bubbles
No odors	Offensive odor	Offensive odor	Decreasing odor
Low bacteria count	Increasing bacterial count	High bacterial count	Decreasing bacterial count
Low organic content	High, decreasing organic content	High, decreasing organic content	Lower organic content
High species diversity	Lower species diversity	Low species diversity	Increased number of species
Normal biotic communities	Increased number of individuals per species	Increased number of individuals per species	Decreased number of individuals
Sensitive organisms present: stone fly nymphs, bass, bluegill, perch, crayfish	Tolerant species: blue green algae, spirogyra, gomphonema, sludgeworms, back swimmers, water boatmen, dragon-flies, gar, carp, catfish,	Pollution adapted species: blue green algae, sludgeworms, mosquito larvae, air breathing snails, no fish.	Tolerant species, some clean water types: blue green algae, euglena, phlorophytes chlorella, spirogyra, blackfly larvae, giant water bugs, clams, catfish, sunfish
Clean, sludge-free bottom	Slime molds and sludge deposits on bottom	Bottom slime blanket, floating sludge	Less bottom slime, some sludge deposits

Adapted from Spellman, Frank R., *Stream Ecology and Self-Purification*. Lancaster, PA: Technomic Publishing Company, Inc., 1996, pp. 72–74.

clean zone conditions. Depending on the volume of water the stream carries, the amount of pollution, and the speed the stream travels, the return to clean zone conditions can take several miles.

Heavy organic pollution levels place demands on the stream biota to regain that balance. When streams become over-loaded with pollution or non-biodegradable man-made ingredients (contaminants), the stream's natural purification processes cannot return the water to its normal quality. If the stream biota cannot handle the pollutant load, they die off, killing the stream.

1.2.2 GROUNDWATER PURIFICATION

Groundwater is filtered naturally as it percolates through the soil and, as long as normal soil conditions exist, groundwater is generally free of contamination. Groundwater is replenished from a percentage of the water that falls to earth each year. Some of the water from precipitation seeps into the ground, filling every interstice, hollow, and cavity. Water is purified as it flows through the soil, by soil processes that remove many impurities and kill disease organisms. Soil is a filter that effectively removes suspended solids and sediments, leaving uncontaminated groundwater generally safe to drink with minimum treatment.

Groundwater quality is of growing concern in areas where leaching agricultural or industrial pollutants, or substances from leaking underground storage tanks, are contaminating groundwater. Once contaminated, groundwater is often difficult to restore (see Figure 1.3).

1.3 THE URBAN WATER CYCLE

Artificially generated water cycles or *urban water cycles* (see Figure 1.4) use either surface or groundwater sources to meet the water supply needs of their communities. Municipalities distribute treated water to households and industries, then collect the wastewater in a sewer system, carrying it to a treatment plant for processing and eventual disposal. Current processing technologies cannot economically provide complete recovery of the original water quality, and treated wastewater is outfalled into running waters to undergo natural stream purification before it is safe for human use as a water supply downstream.

Hammer and Hammer (1996) point out that indirect water reuse processes (see Figure 1.5) provide an example of an artificial water cycle with the natural hydrologic scheme. This involves

- surface-water withdrawal, processing, and distribution
- wastewater collection, treatment, and disposal with dilution to surface water

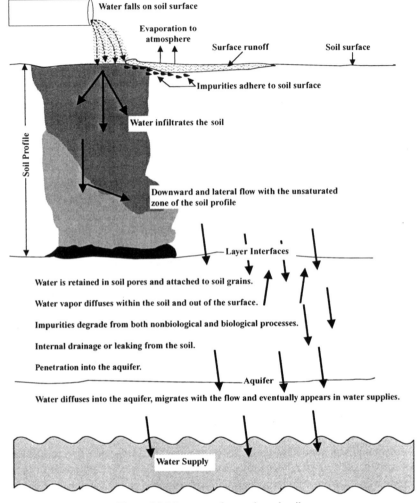

Figure 1.3 Movement of water through soil.

- natural stream self-purification
- cycle repetition by communities downstream (p. 1)

1.4 SAMPLING AND TESTING

Ascertaining water and wastewater conditions, determining how well the sample source meets quality characteristics, and maintaining process efficiency

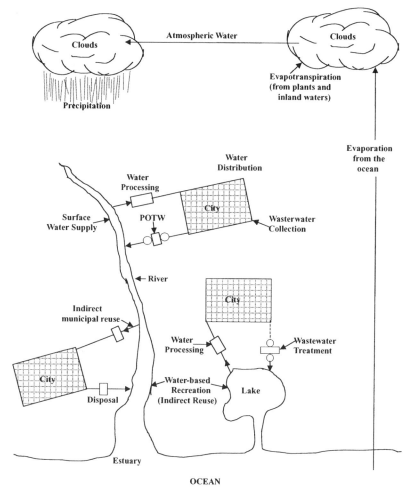

Figure 1.4 Urban water cycle.

all involve obtaining valid sampling and test results. The results of water and wastewater sample testing provide the basis for decisions affecting public health. Critical sampling and testing factors include proper water sampling techniques, proper calibration and use of equipment, and effective sample preparation and preservation (see Photo 1.1).

Whether samples are obtained by grab sampling (a single sample collected over a very short period of time) or composite sampling (obtained by mixing individual grab samples taken at regular intervals over the sampling period), effective handling prevents sample contamination and deterioration prior to

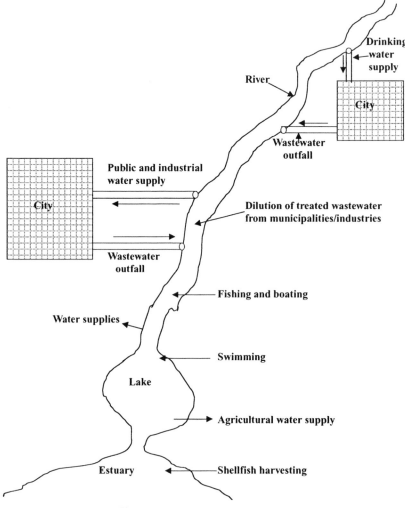

Figure 1.5 Indirect water reuse process.

testing. Accurate and exact testing techniques provide effective data upon which to base treatment processes.

SUMMARY

Ensuring that the water we return to the hydrogeologic cycle after treatment causes no harm to downstream users is of concern to us all—and is a matter

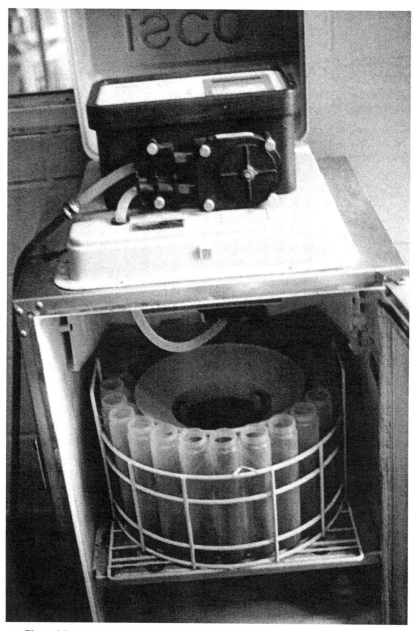

Photo 1.1 Automatic sampler (Lancaster, PA Advanced Wastewater Treatment Plant).

controlled and regulated by federal, state, and local law. We discuss the regulations that govern water quality, and water quality parameters in Chapter 2.

REFERENCE

Hammer, M. J. and M. J. Hammer, Jr. *Water and Wastewater Technology*, 3rd ed. Englewood Cliffs, NJ: Prentice Hall, 1996.

PART II

BASICS OF WATER TREATMENT

Fire Hydrant.

Water Regulations, Parameters, and Characteristics

2.1 PURPOSE: QUALITY PARAMETERS FOR WATER

THE unit processes used to prepare raw or untreated water for public use and consumption are controlled and determined by water quality parameters. These parameters are set by federal regulations and are supported and strengthened by state law. Individual facilities must prove they meet regulatory standards through regulated programs of testing and reporting.

2.2 PURPOSE: WATER TREATMENT

Treatment for drinking water removes from raw water those contaminants (Table 2.1) harmful or unpleasant to humans by a confirmed series of treatment steps or unit processes that produce safe potable water. In raw water treatment, the goals are to remove pollutants that affect water quality and to ensure that water safe for consumption is delivered to the consumer.

2.3 WATER QUALITY: FEDERAL REGULATIONS

Water quality standards are controlled by federal regulation, applied on all levels. After water quality standards became law in the 1970s, the condition of our drinking water supplies improved dramatically. These improvements were the result of two critically important regulations: The Safe Drinking Water Act (SDWA), passed by Congress in 1974, and the Water Pollution Control Act Amendments of 1972 (Clean Water Act, CWA).

Conestoga Water Treatment Plant (Lancaster, PA).

TABLE 2.1. Common Chemical Pollutants.

Source	Common Associated Pollutants
Cropland	Turbidity, phosphorus, nitrates, temperature, total solids
Forestry harvest	Turbidity, temperature, total solids
Grazing land	Fecal bacteria, turbidity, phosphorus
Industrial discharge	Temperature, conductivity, total solids, toxics, pH
Mining	pH, alkalinity, total dissolved solids
Septic systems	Fecal bacteria, (i.e., *Escherichia coli*, enterococcus), nitrates, phosphorus, dissolved oxygen/biochemical oxygen demand, conductivity, temperature
Sewage treatment plants	Dissolved oxygen and BOD, turbidity, conductivity, phosphorus, nitrates, fecal bacteria, temperature, total solids, pH
Construction	Turbidity, temperature, dissolved oxygen and BOD, total solids, toxics
Urban runoff	Turbidity, phosphorus, nitrates, temperature, conductivity, dissolved oxygen, BOD

2.3.1 DRINKING WATER REGULATIONS

The Safe Drinking Water Act of 1974 required the USEPA to establish mandatory drinking water standards for all public water systems serving 25 or more people, or having 15 or more connections. Under the SDWA, the EPA established maximum contaminant levels for drinking water delivered through public water distribution systems. If water analysis indicates a water system is exceeding a maximum contamination level (MCL) for a contaminant, the system must either stop providing the water to the public or treat the water to reduce the contaminant concentration to below the MCL.

Secondary drinking water standards (Table 2.2) are EPA-issued guidelines (and thus, unlike MCLs, are not mandatory) that apply to drinking water contaminants known to adversely affect odor and appearance—water's aesthetic qualities. While these qualities present no known health risk to the public, most drinking water systems comply, if for no other reason than consumer relations. It would be difficult to convince us that the water from our taps is safe if it has an unpleasant smell or is an odd color.

New federal drinking water standards slated to go into effect in February 2001 [The Disinfectant/Disinfection By-Product Rule (D/DBP) and the Interim Enhanced Surface Water Treatment Rule], are aimed at filtering out higher levels of impurities from drinking water and helping communities to upgrade their treatment systems. Meant to simultaneously reduce health threats from bacteria, protozoa, and viruses, as well as from disinfectants, these new amendments toughen standards for allowable concentrations of chlorine by-products in drinking water, regulate *Cryptosporidium*, and tighten standards

TABLE 2.2. National Secondary Drinking Water Standards.

Contaminants	Suggested Levels	Contaminant Effects
Aluminum	0.05–0.2 mg/L	Discoloration of water
Chloride	250 mg/L	Salty taste; corrosion of pipes
Color	15 color units	Visible tint
Copper	1.0 mg/L	Metallic taste; blue-green staining of porcelain
Corrosivity	Noncorrosive	Metallic taste; fixture staining corroded pipes (corrosive water can leach pipe materials, such as lead, into drinking water)
Fluoride	2.0 mg/L	Dental fluorosis (a brownish discoloration of the teeth)
Foaming agents	0.5 mg/L	Aesthetic: frothy, cloudy, bitter taste, odor
Iron	0.3 mg/L	Bitter metallic taste; staining of laundry, rusty color, sediment
Manganese	0.05 mg/L	Taste; staining of laundry, black to brown color, black staining
Odor	3 threshold odor	Rotten egg, musty, or chemical smell
pH	6.5–8.5	Low pH: bitter metallic taste, corrosion. High pH: slippery feel, soda taste, deposits
Silver	0.1 mg/L	Argyria (discoloration of skin), graying of eyes
Sulfate	250 mg/L	Salty taste; laxative effects
Total dissolved solids (TDS)	500 mg/L	Taste and possible relation between low hardness and cardiovascular disease; also an indicator of corrosivity (related to lead levels in water); can damage plumbing and limit effectiveness of soaps and detergents
Zinc	5 mg/L	Metallic taste

Source: USEPA 810-F-94-002, May 1994, p. 6.

for turbidity from 5 μm to 1 μm. Estimates suggest that compliance with these standards will reduce individual exposure to disinfection by-products [including trihalomethanes (THMs)] by 25% (see Appendix A).

2.4 WATER QUALITY CHARACTERISTICS

Water quality parameters provide a yardstick by which to measure water's physical, chemical, and biological characteristics. These parameters include a range of characteristics that make water appealing and useful to consumers,

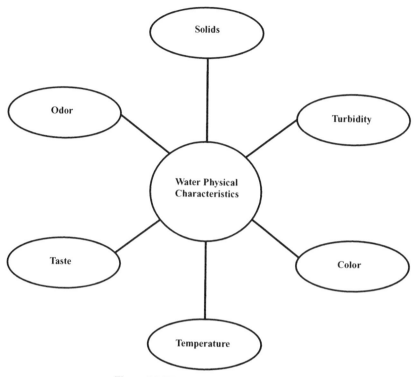

Figure 2.1 Water's physical characteristics.

and that ensure the water presents no harm or disruption to the environment or to humans within a wide range of possible water uses.

2.5 PHYSICAL WATER QUALITY CHARACTERISTICS

Water's physical characteristics (those detectable by sight, touch, taste, or smell) include suspended solids, turbidity, color, temperature, taste and odor (see Figure 2.1).

2.5.1 SOLIDS IN WATER

Solids removal is of great concern in drinking water treatment. Suspended materials provide adsorption sites for biological and chemical agents, and give microorganisms protection against chlorine disinfectants. As suspended solids degrade biologically, they can create objectionable by-products.

Solids can be either suspended or dissolved in water, and are classified by their size and state, by their chemical characteristics, and by their size distribution.

These solids consist of inorganic or organic particles, or of immiscible liquids such as oils and greases. Surface waters often contain inorganic solids such as clay, silt, and other soil constituents as the result of erosion. Organic materials (including plant fibers and biological solids such as bacteria) are also common in surface waters. Groundwater seldom contains suspended solids because of soil's filtering properties.

Filtration provides the most effective means of removing solids in water treatment, although colloids and some other dissolved solids cannot be removed by filtration.

2.5.2 TURBIDITY

Water's clarity is usually measured against a turbidity index. Insoluble particulates scatter and absorb light rays, impeding the passage of light through water. Turbidity indices measure light passage interference (see Figure 2.2). The index starts with 1, showing little or no turbidity, to 5, allowing no passage of light.

Surface water turbidity can result from very small particulate colloidal material (rock fragments, silt, clay, and metal oxides from soil) contributed by erosion or by microorganisms and vegetable materials.

2.5.3 COLOR

Pure water has no color, but foreign substances can often tint water. These include organic matter from soils, vegetation, minerals and aquatic organisms, or municipal and industrial wastes. Color can be a treatment problem that exerts chlorine demand, reducing chlorine's disinfectant effectiveness.

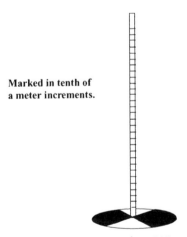

Marked in tenth of
a meter increments.

Figure 2.2 Secchi disk measures turbidity.

Water's color is classified as true color (from dissolved solids that remain after suspended matter is removed) or apparent color (from suspended matter). True color is the most difficult to remove. Measured by comparing a water sample with standard color solutions or colored glass disks, one color unit is equal to the color produced by a 1 mg/L solution of platinum.

Water's color affects its marketability for domestic and industrial use. While colored water does not present a safety issue, most people object to obviously colored water. Colored water can affect laundering, food processing, papermaking, manufacturing, and textiles.

2.5.4 TASTE AND ODOR

Water's taste and odor (the terms are used jointly when used to describe drinking water) is another aesthetically important issue with little safety impact. Taste and odor problems can be caused by minerals, metals, and salts from the soil, and wastewater constituents, and biological reaction end products.

A common method to remove taste and odor from drinking water is to oxidize the problem materials with potassium permanganate, chlorine, or other oxidant. Powdered activated carbon is also used for taste and odor control.

2.5.5 TEMPERATURE

Surface water and groundwater temperature are affected both naturally and artificially. Heat or temperature change in surface waters affects the solubility of oxygen in water, the rate of bacterial activity, and the rate at which gases are transferred to and from water, as well as the health of the fish population.

Temperature is one of the most important parameters in natural surface water systems because such systems are subject to great temperature variations; other than this, temperature is not commonly used for water quality evaluation.

Water temperature does, however, affect in part the efficiency of some water treatment processes. Temperature affects chemical dissolve and reaction rates. Cold water requires more chemicals for efficient coagulation and flocculation. High water temperatures can increase chlorine demand because of increased reactivity, as well as increased levels of algae and other organic matter in the raw water.

2.6 CHEMICAL WATER QUALITY CHARACTERISTICS

The major chemical parameters of concern in water treatment are total dissolved solids (TDS), alkalinity, hardness, fluorides, metals, organics and nutrients, pH, and chlorides. The solvent capabilities of water are directly related to its chemical parameters (see Figure 2.3).

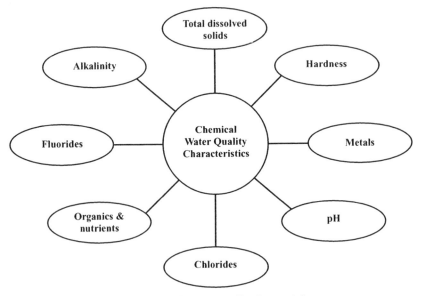

Figure 2.3 Chemical water quality characteristics.

2.6.1 TOTAL DISSOLVED SOLIDS

Solids in water occur either in solution or in suspension. The solids in the water that remain after filtration and evaporation as residue are called total dissolved solids, or TDS. Dissolved solids can be removed from water by filtration and evaporation, and also by electrodialysis, reverse osmosis, or ion exchange.

Dissolved solids may be organic or inorganic, and come from water's contact with substances in soil, on surfaces, and in the atmosphere. Organic dissolved constituents come from decayed vegetation, and from organic chemicals and gases. These dissolved minerals, gases, and organic constituents may cause physiological effects, as well as color, taste, and odor problems.

2.6.2 ALKALINITY

Alkalinity is an important water quality parameter, because it measures the water's ability to neutralize acids. Alkalinity constituents in natural water supplies are bicarbonate, carbonate, and hydroxyl ions—mostly the carbonates and bicarbonates of sodium, potassium, magnesium, and calcium. Alkalinity also occurs naturally from carbon dioxide (from the atmosphere and as a by-product of microbial decomposition of organic material) and from mineral origins (primarily from chemical compounds dissolved from rocks and soil).

While highly alkaline waters do not seriously affect human health, elevated alkalinity can cause an objectionable bitter taste. In treatment, alkaline water can cause problems with the reactions that occur between alkalinity and certain substances in the water that can foul water system appurtenances.

2.6.3 HARDNESS

Hardness in water usually indicates the presence of such minerals as calcium and magnesium. These dissolved minerals cause scale deposits in hot water pipes and affect soap efficiency. These problems make hard water generally unacceptable to the public, though advantages to hard water do exist. Hard water helps tooth and bone growth and hard water scaling reduces toxicity of lead oxide in pipelines made of lead.

2.6.4 FLUORIDES

Fluoride is toxic to humans in large quantities, and to some animals, though moderate amounts of fluoride ions (F^-) in drinking water contribute to good dental health. Fluoride appears in groundwater in only a few geographical regions, and in a few types of igneous or sedimentary rocks. It is seldom found in appreciable quantities in surface water. Fluoride is a common addition to drinking water in many communities.

2.6.5 METALS

Metals in water that are harmful in relatively small amounts are classified as toxic; other metals are classified as nontoxic. In natural waters other than groundwater, metal sources include dissolution from natural deposits and discharges of domestic, agricultural, or industrial wastes. Leachate from improperly designed, constructed, or managed landfills is another common source.

Some metals [iron (Fe) and manganese (Mn), for example] impart a bitter taste to drinking water even at low concentrations, though they do not cause health problems. These metals usually occur in groundwater in solution. They and other metals in solution may cause brown or black stains on laundry and on plumbing fixtures.

2.6.6 ORGANICS

Organic matter in water can cause color problems as well as taste and odor problems. Organic matter can contribute to the formation of halogenated compounds in water undergoing chlorine disinfection. Organic matter can also create problems with oxygen depletion in streams, because as microbes metabolize organic material, they consume oxygen. Oxygen depletion from organic matter interferes with water treatment processes (see Photo 2.1).

Photo 2.1 Natural surface waters may contain high levels of organics that may cause color, taste, or odor problems (Blackwater Falls, WV).

30

The oxygen microbes consume is dissolved oxygen (DO). This demand for oxygen is called the biochemical oxygen demand (BOD), the amount of dissolved oxygen aerobic decomposers require to decay organic materials in a given volume of water over a five-day incubation period at 68°F (20°C). If the oxygen is not continually replaced, the DO level decreases as the microbes decompose the organics, until the cycle fails from lack of available oxygen.

Generally, organic matter in water comes from natural sources—decaying leaves, weeds, and trees. Man-made sources include pesticides and other synthetic organic compounds (see Table 2.3).

Many organic compounds are soluble in water, and surface waters are more prone to contamination by natural organic compounds than are groundwaters. Dissolved organics are usually divided into biodegradable and nonbiodegradable.

Nonbiodegradable organics resist biological degradation. For example, the refractory (resistant to biodegradation) constituents of woody plants (tannin and lignic acids, phenols, and cellulose) are found in natural water systems. Other essentially nonbiodegradable constituents include some polysaccharides with exceptionally strong bonds, and benzene with its ringed structure (associated with the refining of petroleum).

2.6.7 INORGANICS

Natural water can possess several common inorganic constituents (pH, chlorides, alkalinity, nitrogen, phosphorous, sulfur, toxic inorganic compounds, and heavy metals) that are important to treatment processes. Water's inorganic load is affected by wastewater discharges, geologic conditions and formations, and inorganics that remain in water after evaporation (Snoeyink and Jenkins, 1988). Natural waters dissolve rocks and minerals, adding inorganics to water, and these constituents can also enter water through human use (see Table 2.4).

Inorganic contaminants are often removed by corrosion control methods or by removal techniques. Corrosion controls reduce corrosion by-products (lead, for example) in potable water. Removal technologies, including coagulation/filtration, reverse osmosis (RO), and ion exchange, are used to treat source water contaminated with metals or radioactive substances.

2.6.7.1 Nutrients

The nutrients of greatest concern in water supplies are nitrogen and phosphorous. Other nutrients include carbon, sulfur, calcium, iron, potassium, manganese, cobalt, and boron—all essential to the growth and reproduction of plants and animals.

TABLE 2.3. Primary Standard MCLs and MCLGs for Organic Chemicals.

Contaminant	Health Effects	MCL—MCLG (mg/L)	Sources
Aldicarb	Nervous system effects	0.003—0.001	Insecticide
Benzene	Possible cancer risk	0.005—zero	Industrial chemicals, paints, plastics, pesticides
Carbon tetrachloride	Possible cancer risk	0.005—Zero	Cleaning agents, industrial wastes
Chlordane	Possible cancer	0.002—Zero	Insecticide
Endrin	Nervous system, liver, kidney effects	0.002—0.002	Insecticide
Heptachlor	Possible cancer	0.0004—Zero	Insecticide
Lindane	Nervous system, liver, kidney effects	0.0002—0.0002	Insecticide
Pentachloro-phenol	Possible cancer risk, liver, kidney effects	0.001—Zero	Wood preservative
Styrene	Liver, nervous system effects	0.1—0.1	Plastics, rubber, drug industry
Toluene	Kidney, nervous system, liver, circulatory effects	1—1	Industrial solvent, gasoline additive, chemical manufacturing
Total trihalomethanes (TTHM)	Possible cancer risk	0.1—Zero	Chloroform, drinking water chlorination by-product
Trichloro-ethylene (TCE)	Possible cancer risk	0.005—Zero	Waste from disposal of dry cleaning material and manufacture of pesticides, paints, waxes; metal degreaser
Vinyl chloride	Possible cancer	0.002—Zero	May leach from PVC pipe
Xylene	Liver, kidney, nervous system effects	10—10	Gasoline refining by-product, paint ink, detergent

Source: USEPA 810-F-94-002, May 1994.

32

TABLE 2.4. Primary Standard MCLs for Inorganic Chemicals.

Contaminant	Health Effects	MCL (mg/L)	Sources
Arsenic	Nervous system effects	0.05	Geological, pesticide residues, industrial waste, smelter operations
Asbestos	Possible cancer	7 MFL[a]	Natural mineral deposits, A/C pipe
Barium	Circulatory system effects	2	Natural mineral deposits, paint
Cadmium	Kidney effects	0.005	Natural mineral deposits, metal finishing
Chromium	Liver, kidney, digestive system effects	0.1	Natural mineral deposits, metal finishing, textile and leather industries
Copper	Digestive system effects	TT[b]	Corrosion of household plumbing, natural deposits, wood preservatives
Cyanide	Nervous system effects	0.2	Electroplating, steel, plastics, fertilizer
Fluoride	Dental fluorosis, skeletal effects	4	Geological deposits, drinking water additive, aluminum industries
Lead	Nervous system and kidney effects, toxic to infants	TT	Corrosion of lead service lines and fixtures
Mercury	Kidney, nervous system effects	0.002	Industrial manufacturing, fungicide, natural mineral deposits
Nickel	Heart, liver effects	0.1	Electroplating, batteries, metal alloys
Nitrate	Blue-baby effect	10	Fertilizers, sewage, soil & mineral deposits
Selenium	Liver effects	0.05	Natural deposits, mining, smelting

[a] Million fibers per liter.
[b] Treatment techniques have been set for lead and copper because the occurrence of these chemicals in drinking water usually results from corrosion of plumbing materials. All systems that do not meet the *action level* at the tap are required to improve corrosion control treatment to reduce the levels. The action level for lead is 0.015 mg/L, and for copper it is 1.3 mg/L.
Source: EPA 810-94-002, May 1994.

Nitrogen (N_2) is an extremely stable gas. As the primary component of the earth's atmosphere, it occurs in many forms in the environment and takes part in many biochemical reactions. Nitrogen enters water from runoff from animal feedlots, fertilizer runoff, municipal wastewater discharges, and from certain bacteria and blue-green algae that directly obtain atmospheric nitrogen. Some types of acid rain also add nitrogen to surface waters.

In water, nitrogen in the form of nitrate (NO_3) indicates contamination with sewage. An immediate health threat to both human and animal infants, excessive nitrate concentrations in drinking water can even cause death.

Though the presence of phosphorous (P) in drinking water has little effect on health, too much phosphorus in water supplies causes problems. While essential for growth, excess amounts of this nutrient contribute to algae bloom and lake eutrophication. Phosphorous sources include phosphates from detergents, fertilizer and feedlot runoff, as well as municipal wastewater discharges.

2.6.7.2   pH

pH (hydrogen ion concentration) indicates the intensity of acidity or alkalinity in water, and affects biological and chemical reactions. Water's chemical balance (equilibrium relationships) is strongly influenced by pH. For example, water's pH levels directly affect certain unit processes, including disinfection with chlorine. Increased pH increases the contact time needed for chlorine disinfection.

2.6.7.3   Chlorides

Chloride (a major inorganic constituent in water) generally does not cause any harmful effects to public health, though a high enough concentration can cause an objectionable salty taste. Chlorides occur naturally in groundwater,

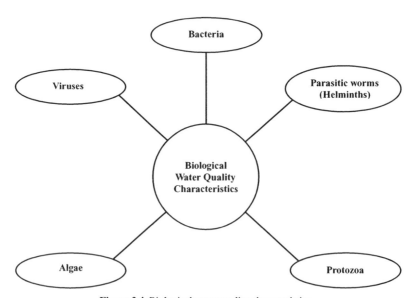

Figure 2.4 Biological water quality characteristics.

streams, and lakes, but concentrations in fresh water of 500 mg/L or more may indicate sewage contamination.

2.7 BIOLOGICAL WATER QUALITY CHARACTERISTICS

Whether or not living organisms are present in water is a very useful indicator of water quality. Thousands of biological species spend part, if not all, of their life cycles in water. All members of the biological community can provide water quality parameters (see Figure 2.4).

Most water-borne microbes are beneficial, particularly as food chain decomposers. Only a few microorganism species cause disease in humans or damage to the environment: pathogens are organisms capable of infecting or transmitting diseases to humans and animals (Table 2.5).

The presence or absence of pathogens in water is of primary importance. Pathogens include species of bacteria, viruses, algae, protozoa, and parasitic worms (helminths). Although they do not naturally occur in aquatic environments, pathogens can be transmitted by natural water systems.

TABLE 2.5. Waterborne Disease-Causing Organisms.

Microorganism	Disease
Bacterial	
Escherichia coli	
Salmonella typhi	Typhoid fever
Salmonella sp.	Salmonellosis
Shigella sp.	Shigellosis
Yersinia entercolitica	Yersiniosis
Vibrio cholerae	Cholera
Campylobacter jejuni	Campylobacter enteritis
Legionella	Legionellosis
Intestinal Parasites	
Entamoeba histolytica	Amebic dysentery
Giardia lamblia	Giardiasis
Cryptosporidium	Cryptosporidiosis
Viral	
Norwalk agent	—
Rotavirus	—
Enterovirus	Polio
	Aseptic meningitis
	Herpangina
Hepatitis A	Infectious hepatitis
Adenoviruses	Respiratory disease
	Conjunctivitis

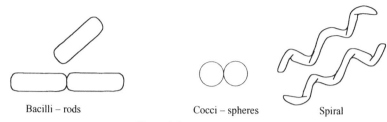

Bacilli – rods   Cocci – spheres   Spiral

Figure 2.5 Bacterial shapes.

2.7.1 BACTERIA

Air, water, soil, rotting vegetation, and human and animal intestines all contain bacteria. While most bacteria we encounter are harmless, waterborne pathogenic bacteria transmit diseases that cause common symptoms of gastrointestinal disorder. Eliminating pathogenic organisms through chemical treatment ensures safe drinking water to the consumer (see Figure 2.5).

2.7.2 VIRUSES

Viruses are tiny entities that require a host to live and reproduce. They carry the information they need for replication, but not the required machinery. The host provides that machinery. Waterborne viral infections generally cause nervous system disorders, not gastrointestinal ones.

Because the many varieties of viruses are small in size, unstable in behavior and appearance, and occur in low concentrations in natural waters, testing for viruses in water is difficult (see Figure 2.6). This difficulty is compounded by

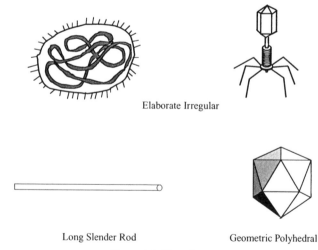

Elaborate Irregular

Long Slender Rod   Geometric Polyhedral

Figure 2.6 Virus shapes.

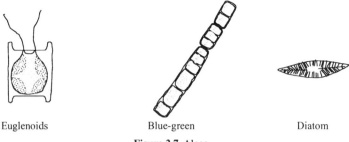

| Euglenoids | Blue-green | Diatom |

Figure 2.7 Algae.

limited identification methods. Add to this concern about disinfection effectiveness, and the reasons that viruses are of special concern in water treatment become apparent.

2.7.3 ALGAE

Algae are microscopic plants that occur in fresh water, salt water, polluted water, and wastewater. Since most need sunlight to live, they only grow where they can be exposed to light—near the water surface (see Figure 2.7).

Algae play an important role in lake eutrophication (aging). In general, algae are considered nuisance organisms because they create taste and odor problems in public water supplies, and because removing them from the water causes extra expense.

2.7.4 PROTOZOA

Protozoa, the simplest of animal species, are mobile, single-celled, completely self-contained organisms (see Figure 2.8). Some protozoa are free-living—others (only a few) are parasitic. They can be pathogenic or non-pathogenic, microscopic or macroscopic. Highly adaptable, protozoa are widely distributed in natural waters. Most protozoa are harmless; only a few cause illness in humans—*Entamoeba histolytica* (amebiasis), *Cryptosporidium parvum*, and *Giardia lamblia* (giardiasis) are three important exceptions (see Figures 2.9 and 2.10). The cysts aquatic protozoans form during adverse environmental conditions make them difficult to deactivate by disinfection. Filtration is usually the most effective means to remove them from water.

2.7.5 WORMS (HELMINTHS)

Worms inhabit organic mud and slime. They have aerobic requirements and can metabolize solid organic matter that other microbes cannot degrade. Human and animal wastes containing worms usually seed supply waters with worms.

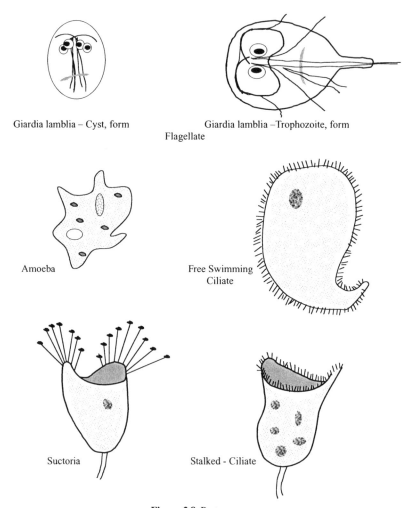

Giardia lamblia – Cyst, form   Giardia lamblia –Trophozoite, form
                                             Flagellate

Amoeba   Free Swimming
                                 Ciliate

Suctoria   Stalked - Ciliate

Figure 2.8 Protozoa.

Worms are a hazard primarily to those who directly contact untreated water. *Tubifix* worms are common indicator organisms for pollution in streams.

2.7.6 INDICATOR ORGANISMS

Because of the difficulties inherent in microbial testing and identification, a method for obtaining a clear indication of a source water's condition (the presence or absence of pollution) is essential. Water treatment common practice does not test for individual pathogens (which may be present in such small quantities that they are undetectable), which would mean testing and

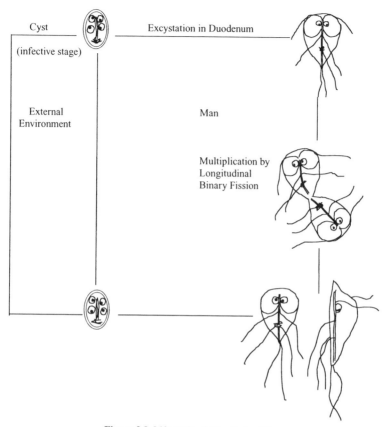

Figure 2.9 Life cycle of *Giardia lamblia*.

retesting for each pathogenic organism—an enormously time-consuming (and thus costly) process. Instead, common practice requires testing for a single species of indicator organism. A positive test for the indicator organism alerts us to the possible presence of sewage contamination. The indicator organisms used are the coliforms. When coliforms are present in a water sample, they indicate sewage contamination. A water source contaminated with sewage may also contain pathogenic microorganisms and presents a threat to public health.

2.7.7 COLIFORMS

Coliform group testing indicates a proportion of contamination relative to an easily defined quantity of water. It involves estimating the number of fecal coliform bacteria present in a measured water sample. Not only are fecal coliforms

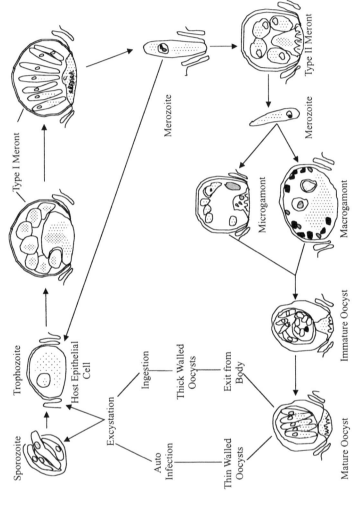

Figure 2.10 Life cycle of *Cryptosporidium parvum.*

Type I Meront

Merozoite

Type II Meront

Merozoite

Merozoite

Microgamont

Macrogamont

Trophozoite

Immature Oocyst

Sporozoite

Host Epithelial Cell

Ingestion

Thick Walled Oocysts

Exit from Body

Excystation

Mature Oocyst

Auto Infection

Thin Walled Oocysts

always present in fecal wastes, water recently contaminated with sewage will always contain coliforms, which will outnumber disease-producing organisms.

Fecal coliform bacteria do not themselves cause disease. These organisms are present in the intestinal tract of all mammals. Human bodily wastes contain literally millions of coliforms. While the correlation between coliforms and human pathogens in natural waters is not absolute, the number of fecal coliform bacteria present effectively indicates the water source's pollution levels.

SUMMARY

A water source's normal condition is compared to the standards, and taking the raw water influent to accepted standards is the foundational task of water treatment. The parameters and testing methods lay the groundwork for the processes used in effective, efficient water treatment.

REFERENCES

Snoeyink, V. L. and D. Jenkins. *Water Chemistry*, 2nd ed. New York: John Wiley and Sons, 1988.

USEPA. *Selected Primary MCSs & MCLGs for Organic Chemicals*. Washington, D.C., USEPA 810-F-94-002, May 1994.

CHAPTER 3

Water Purification: System Overview

3.1 PROCESS PURPOSE

W ATER treatment brings raw water up to drinking water quality. The processes this entails depend on the quality of the water source. Surface water sources (lakes, river, reservoirs, and impoundments) generally require higher levels of treatment than groundwater sources. Groundwater sources may incur higher operating costs from machinery, but may require only simple disinfection (see Figure 3.1).

3.2 WATER TREATMENT UNIT PROCESSES

Treatment for raw water taken from groundwater and surface water supplies differs somewhat, but one commonly employed treatment technology illustrates many of the unit processes involved. Primary treatment processes for surface water supplies include the basic water treatment processes shown in Table 3.1 and hardness removal (not mandatory), as well as:

- *Intake* to bring in for treatment water of the best possible quality the source can provide.
- *Screening* to remove floating and suspended debris of a certain size.
- *Chemical mixing* with the water to allow suspended solids to coagulate into larger particles that settle more easily.
- *Coagulation*, a chemical water treatment method that causes small particles to stick together to form larger particles.
- *Flocculation* to gently mix the coagulant and water, encouraging large floc particle formation.
- *Sedimentation* to slow the flow so that gravity settles the floc.

Lake Redman, York County, PA.

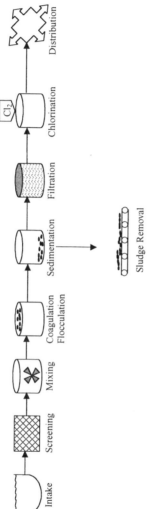

Figure 3.1 Unit processes: Water treatment.

45

TABLE 3.1. Basic Water Treatment Processes.

Process/Step	Purpose
Intake	Conveys water from source to treatment plant
Screening	Removes large debris (leaves, sticks, fish) that could foul or damage plant equipment
Chemical pretreatment	Conditions the water for removing algae and other aquatic nuisances
Presedimentation	Removes gravel, sand, silt, and other gritty materials
Microstraining	Removes algae, aquatic plants and remaining debris
Chemical feed and rapid mix	Adds chemicals—coagulants, pH adjusters, etc.
Coagulation/flocculation	Converts nonsettleable or settleable particles
Sedimentation	Removes settleable particles
Softening	Removes hardness-causing chemicals
Filtration	Removes particles of solid matter, including biological contamination and turbidity
Disinfection	Kills disease-causing organisms
Adsorption using granular activated carbon (GAC)	Removes radon and many organic chemicals, including pesticides, solvents, and trihalomethanes (THMs)
Aeration	Removes volatile organic chemicals (VOCs), radon, H_2S, and other dissolved gases; oxidizes iron and manganese
Corrosion control	Prevents scaling and corrosion
Reverse osmosis, electrodialysis	Removes nearly all inorganic contaminants
Ion exchange	Removes some inorganic contaminants including hardness-causing chemicals
Activated alumina	Removes some inorganic contamination
Oxidation filtration	Removes some inorganic contaminants—iron, manganese, radium, etc.

Source: Adapted from AWWA, *Introduction to Water Treatment*, Vol. 2, 1984.

- *Sludge processing* to remove the solids and liquids collected in the settling tank, and to dewater and dispose of them.
- *Disinfection* to ensure the water contains no harmful pathogens.

Once water from the source has entered the plant as influent, water treatment processes break down into two parts. The first part, *clarification*, consists of screening, coagulation, flocculation, sedimentation, and filtration. Clarification processes go far in potable water production, but while they do remove many microorganisms from the raw water, they cannot produce water free of microbial pathogens. The final step, *disinfection*, destroys or inactivates disease-causing infection agents.

SUMMARY

The major unit processes that make up the standard water treatment process are presented in the model in Figure 3.1. It shows the water source (in this case a river, since surface water treatment requires more process steps than does groundwater), screening, coagulation, flocculation, sedimentation, filtration, disinfection, and distribution. The remaining chapters in Part 2 cover these processes step-by-step.

REFERENCE

AWWA. *Introduction to Water Treatment*, Vol. 2. Denver: American Water Works Association, 1984.

CHAPTER 4

Sources, Intake, and Screening

4.1 INTRODUCTION: WATER SOURCES

WHILE all communities are different, one element they all have in common is the need for water for industries, commercial enterprises, and residents to use. In fact, a regular supply of potable water is the most important factor affecting whether or not any living creature chooses an area to live. However, stable and plentiful fresh water sources are not always readily available where they could be put to most practical use. We recognize that water is not uniformly distributed. The heaviest populations of any life forms, including humans, are found in regions of the world where potable water is present, because lands barren of water simply won't support large populations. The hydrologic cycle constantly renews our freshwater supplies, but population pressure is constantly increasing. As our population grows, and we move into lands without ready freshwater supplies, we place ecological strain upon those areas, and on their ability to support life.

Communities must have a constant adequate water supply to survive. Populations that chose to build in areas without an adequate local water supply are at risk when emergencies occur. While attention to water source remediation, pollution control, water reclamation and reuse can help to ease the strain increasing populations place on a water supply, technology can't create local freshwater supplies, whether from surface or groundwater sources, it can only stretch the supplies already on hand (Photo 4.1).

4.2 WATER SOURCES

Water treatment brings raw water up to drinking water quality. The processes

49

Water pumping station.

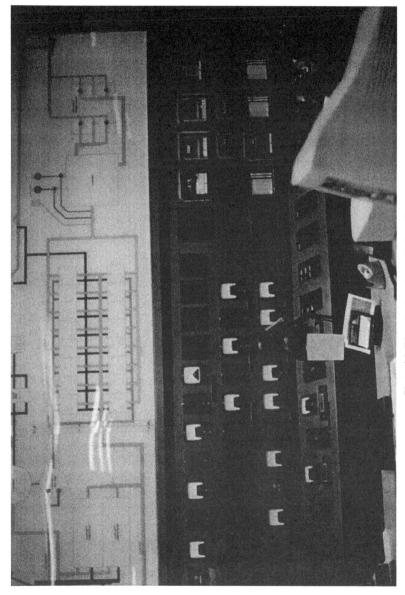

Photo 4.1 Water treatment operator's station.

this entails depend on the quality of the water source—either surface water or groundwater. Common surface water sources include lakes and rivers, and precipitation or spring water that has been channeled into surface water storage in reservoirs. Most drinking water used by larger communities (especially in cities) is taken from surface sources. Surface water sources generally need both filtration and disinfection to be potable (see Photo 4.2).

4.2.1 SURFACE WATER SUPPLIES

In general, surface water is the result of precipitation—either rainfall or snow. An average of about 4,250 billion gallons per day falls on the U.S. mainland, of which about 66% returns to the atmosphere directly from lake and river surface evaporation and plant transpiration. About 1,250 billion gallons per day remain, flowing over or through the earth on their constant travel through the hydrogeologic cycle to return to the sea.

Surface water sources include:

- rivers and streams
- lakes
- impoundments (man-made lakes created by damming)
- shallow wells directly affected by precipitation
- springs whose flow or quantity directly depends on precipitation
- rain catchments (drainage basins)
- tundra ponds or muskegs (peat bogs)

As a source for potable water, surface water has some advantages over groundwater. Surface water is comparatively easy to locate without the aid of a geologist or hydrologist. Surface water is usually not tainted with minerals leached out during contact with the earth.

However, surface water sources are easily contaminated with microorganisms that can cause waterborne diseases, and with chemicals that enter from surrounding runoff and upstream discharges. Especially in areas where surface waters are in short supply, water rights can present problems (see Photo 4.3).

Human interference (influences) and natural conditions affect surface water runoff flow rates. When surface water runs quickly off land surfaces, the water does not have enough time to infiltrate the ground and recharge groundwater aquifers, and can cause erosion and flooding problems. Surface water that runs off quickly usually does not have enough contact time to increase in mineral content—about the only positive to be said for this generally damaging condition.

Surface water collects in drainage basins that direct the water through gravity-driven paths to the ocean. Surface water runoff follows the path of least resistance, normally flowing toward a primary watercourse unless some man-made distribution system diverts the flow. Drainage basins are generally measured

Photo 4.2 Water source (Lake Redman, York County, PA).

Photo 4.3 Water supply protection sign.

in square miles, acres, or sections. A community drawing water from a surface water source must consider drainage basin size.

Several factors affect runoff over land surfaces. These include:

- *Rainfall duration:* any rain, if it lasts long enough, will eventually saturate the soil and allow runoff to take place.
- *Rainfall intensity:* hard-driving rain saturates the soil more quickly than gentle rain. Saturated soil holds no more water; excess water builds up on the surface, creating surface runoff.
- *Soil moisture:* already saturated soil causes surface runoff to occur sooner than dry soil. Frozen soil is basically impervious; snow melt or rain runoff can be 100% off frozen ground.
- *Soil composition:* surface soil composition directly affects runoff amounts. For example, hard rock surfaces result in 100% runoff. Clay soils have small void spaces that close when wet, and do not allow infiltration. Large void spaces in coarse sand allow water to pass easily, even in a torrential downpour.
- *Vegetation cover:* groundcover limits runoff, creating a porous layer (a sheet of decaying natural organic substances) above the soil. This porous "organic" sheet passes water easily into the soil, while acting as a protective cover for the soil against hard, driving rains. Downpours can compact bare soils, close off void spaces, and increase runoff. Vegetation and groundcover maintain the soil's infiltration and water-retention capacity and work to reduce soil moisture evaporation.
- *Ground slope:* up to 80% of rainfall that lands on steeply sloping ground will become surface runoff. Gravity carries the water down the surface more quickly than it can infiltrate. A higher percentage of water flow on flat lands infiltrates the ground; water's movement over the surface is usually slow enough to provide opportunity for higher amounts of infiltration.
- *Human influences:* many human activities impact surface water runoff, most of which tend to increase the rate of water flow. Though man-made dams generally contain the flow of runoff, practices that include channeling water through canals and ditches carry water along rather than encouraging infiltration. Agricultural activities generally remove groundcover that normally retards runoff. Communities provide ample examples of human impact on runoff. Impervious surfaces (including streets, parking lots, and buildings) greatly increase the amount of runoff from precipitation. These man-made surfaces work to hasten the flow of surface water, and can cause devastating flooding. Bad planning can create areas where even light precipitation can cause local flooding. Besides dictating increased need for stormwater management systems, these surfaces do not allow water to percolate into the soil to recharge groundwater supplies—often a serious problem for a location's water supply.

4.2.2 GROUNDWATER SUPPLY

A gigantic water source forms a reservoir that feeds all the natural fountains and springs that flow on the surface of the earth—groundwater that lies contained in aquifers beneath earth's crust. But how does water travel into the aquifers that lie under earth's surface?

Groundwater is replenished through part of the three feet of water that (on the average) falls to earth each year on every square foot of land. When water falls to earth, it follows three courses. Some runs off directly into rivers and streams (roughly six inches of that three feet), eventually traveling to the sea. Evaporation and transpiration account for another approximately two feet. That last six inches seeps into the ground, entering and filling each hollow and cavity, and trickling down into the water table that lies below the surface.

The water table is usually not level. It often follows the shape of the ground surface. Groundwater flows downhill in the direction of the water table slope. Where the water table intersects low points of the ground, it seeps out into springs, lakes, or streams. Almost all groundwater is in constant motion through the pores and crevices of the aquifer in which it occurs.

When compared to surface water as a source, groundwater does present some advantages. Groundwater is not as easily contaminated as surface water, and is usually lower in bacteriological contamination. Groundwater quality and quantity usually remain stable throughout the year, and are less affected by short-term droughts than surface water. Groundwater is available in most locations in the United States.

Disadvantages include higher operating costs, because groundwater supplies must be pumped to the surface. Once contaminated, groundwater is very difficult to restore, and contamination can be difficult to detect. High mineral levels and increased levels of hardness are common to groundwater supplies, because the water is in contact longer with minerals. Groundwater sources near coastal areas may be subject to saltwater intrusion.

4.2.3 WATERSHED MANAGEMENT PROGRAMS

Surface runoff water, exposed and open to the atmosphere, is water flow that has not yet reached a definite stream channel. When the rate of precipitation exceeds either the rate of interception and evapotranspiration, or the amount of rainfall readily absorbed by the earth's surface, the remainder flows over or just underneath the surface as surface water runoff or overland flow. The total land area that contributes runoff to a stream or river is called a *watershed, drainage basin* or *catchment area.*

Groundwater (usually pumped from wells) often requires only simple disinfection, or at the most, softening (removing calcium and magnesium) before disinfection. In rural areas, the most usual source water is subsurface (usually wells). Both groundwater and surface water supplies depend upon levels of precipitation and runoff, and surface water sources, especially, can vary seasonally—reservoirs for water supply storage help to regulate uneven supply conditions. While groundwater is available throughout most of the U.S., the amount available at particular locations may be limited. Both surface and groundwater sources must consider watershed or catchment area conditions and protection (see Figure 4.1).

Watershed and wellhead management practices and formal programs are increasing public awareness of the need for individuals to act responsibly to protect local water supplies. Programs that provide information on non-point source pollutants and how they enter surface and groundwaters, that identify watershed, wellhead or catchment areas, or that identify points where chemical

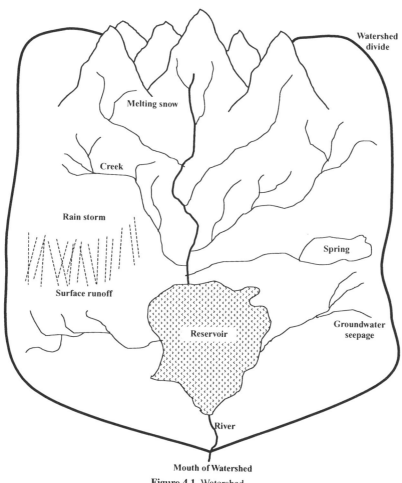

Figure 4.1 Watershed.

dumping will cause damage to water supplies, all bring valuable consumer attention to environmental concerns.

4.3   PROCESS PURPOSE: INTAKE AND SCREENING

Water intake involves bringing the water from the source to the treatment facility (see Figure 4.2). Surface water from rivers, lakes, or reservoirs flows into the transmission system through an intake structure. Groundwater flow moves through an intake pipe from the groundwater source, and is pumped through a transmission conduit to a distribution system. Once the influent enters the plant, surface waters demand more treatment processes than groundwater sources need.

Surface water bodies (whether we use them as a drinking water source or not) unfortunately often contain large quantities of trash, as well as other pollutants. These range from natural pollutants, including brush, branches, logs, rocks, and grit, to the plastic bottles, old paper and cardboard, cigarette butts and packs, lost shoes, and the beer cans people casually toss into the river, to more dangerous and persistent industrial pollutants. In short, the water that a treatment plant may take in for eventual distribution often carries with it a wide range of bulky, long-lasting solids. These solids are physically removed before the water actually enters the plant by screening, the initial treatment step.

Large debris carried along to other treatment processes could damage or foul plant equipment, increase chemical usage, block water flow in pipes or channels, or otherwise slow water treatment.

4.3.1   PROCESS EQUIPMENT: INTAKE

Using water drawn directly from lakes, reservoirs, or rivers means there is a need for selection of the best water possible from the source. Intake structures provide that control over the water supply quality.

4.3.2   SURFACE AND GROUNDWATER WATER INTAKE

Intake structures for lakes, reservoirs, or rivers are in place to accomplish two purposes: to supply water of the best possible quality from the source, and to protect downstream equipment and piping from damage or clogging from debris, flooding, or wave action. Intakes typically consist of screened openings and a conduit to carry flow to a sump, where the water is pumped to the treatment facility. Intake structure must take into account broad differences in water depth, flow, quality, and temperature. Sources of pollution and the direction that current flow will carry them is also of importance, as are navigation, wind and

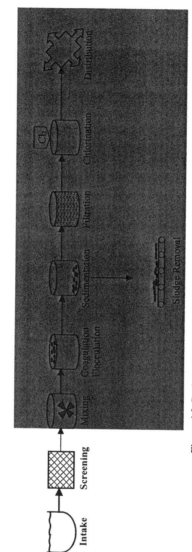

Figure 4.2 Basic water treatment unit processes: water intake and screening.

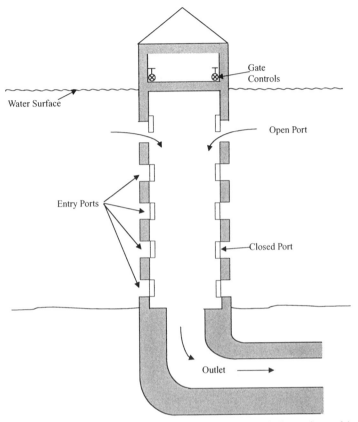

Figure 4.3 Tower water intake for a reservoir or lake water supply (larger than scale).

current patterns, scour and sedimentation deposits, and the amount of floating debris.

Considerations for lake and reservoir intakes are different than for river intakes. Intake types include towers, submerged ports, and shoreline structures (see Figure 4.3 and Photo 4.4). Changes in water level and quality variations with depth are common in lakes and reservoirs, and tower intakes (which provide ports at several depths) are designed for these conditions. In lakes, for most of the year, water quality is best near the surface, but with the changes that occur with spring and fall turnover, and with the possibility of icing in winter, the option of several ports means control over intake selection.

Generally, river intakes are submerged or screened shore intakes, in part because of low costs (see Figure 4.4). River intakes are designed to withdraw water from slightly below the surface. This generally provides the best possible river water quality, by avoiding sediment in suspension in the depths, as well as floating debris.

Photo 4.4 River water intake.

Water Surface

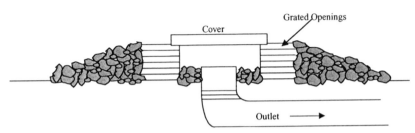

Figure 4.4 Submerged intake used for both lake and river sources.

Groundwater is pumped to an intake pipe from the groundwater source. Pumps carry the water through a transmission conduit to the distribution system.

4.4 PROCESS EQUIPMENT: SCREENING

Removing the floating debris is an important consideration for surface water treatment. Treatment plants use a variety of screening devices to remove the trash and natural debris surface waters carry. Once influent is past the intake, trash screens or rakes, traveling water screens, drum screens, bar screens, or passive screens are employed to remove debris from influent. Screen opening size and flow rate determine the minimum size of debris the screening equipment can halt. Considerations for equipment selection include costs related to operation and equipment, plant hydraulics, debris handling requirements, operator qualifications, and availability (see Photo 4.5).

4.4.1 TRASH SCREENS (RAKES)

Rough and large debris is caught and retained on trash racks by trash screens or trash rakes. Trash screens or rakes are commonly used as a preliminary screening device to remove the largest debris before the influent enters finer screening systems—drum screens or traveling water screens.

Commonly, trash screens are constructed of steel, though high-density polymers are now attracting attention. Trash rack bar spacing ranges from 1.5 to 4 inches, and one or more stationary racks may be used on a trash screen, along with a screen raking system. Raking mechanisms can be installed on dam walls or on the sides of buildings, depending on intake configuration requirements.

Photo 4.5 Debris screened from water intake.

Typically mounted on fixed structures designed to clean a single trash rack, rakes can also be suspended from overhead gantries, or wheel-mounted to cover the complete intake structure width and clean individual sections of a wide trash rack.

4.4.2 TRAVELING WATER SCREENS

Traveling water screens consist of a continuous series of wire mesh panels. These are bolted to basket frames or trays, and attached to two aligned strands of roller chain. Placed in a channel of flowing water to remove floating or suspended debris (see Figure 4.5 and Photo 4.6), traveling water screens run on a sprocket assembly to cover a vertical path through the flow. The raw water passes through the revolving baskets, while the debris load is carried out of the stream flow and above the screen operating level, where high-pressure water sprays remove it. Some traveling water screens operate continuously, others intermittently. Intermittent water screens are activated by set time intervals or by metered specific headloss levels.

In river installations in particular, debris load, water depth, and water flow conditions can widely fluctuate. An individual water screen installation should consider wire mesh size, influent velocity through the mesh, basket or channel width, the number of screens, the maximum average flow, maximum/minimum average water levels, and the type of service, as well as the starting/operating headloss requirements.

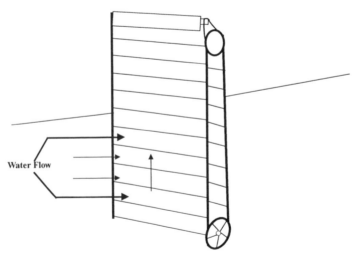

Figure 4.5 Traveling water screen.

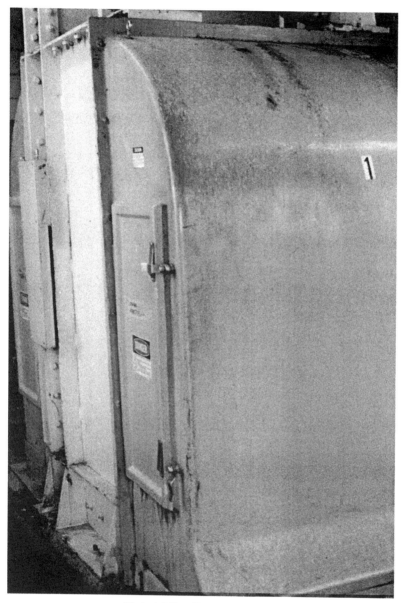

Photo 4.6 Traveling screen housing.

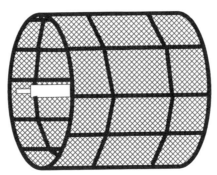

Figure 4.6 Drum screen.

4.4.3 DRUM SCREENS

Drum screens have very few moving parts. Simplicity of design and construction keep installation, operation, and maintenance costs relatively low. The screen itself (a series of wire mesh panels attached to the periphery of a cylinder) is mounted on a horizontal axis (see Figure 4.6). The screen turns slowly on its axis, picking up debris as the water flows through it.

4.4.4 BAR SCREENS

Bar screens are used at some water treatment facilities, though they are more common at wastewater facilities. Designed to handle relatively large debris, a bar screen consists of a rack of straight steel bars welded at both ends to horizontal steel members. Powered rakes move up and down the bar rack face (see Figure 4.7 and Photo 4.7) removing debris, and elevating in and out of flow. The debris is removed at the top of the operating cycle by a wiper mechanism. Bar screen assemblies are normally installed at a 60 to 80° angle from the horizontal.

4.4.5 PASSIVE SCREENS

The most mechanically simple screening devices are passive intake screens or stationary screening cylinders (see Figure 4.8). They have no moving parts, and careful placement ensures they use no debris moving or handling equipment. Usually, passive intake screens are mounted on a horizontal axis, and are oriented parallel to the current in a surface water body to take advantage of natural ambient currents. The current action works positively to clean the screen. Controlled through-screen velocities (a maximum intake velocity of 0.5 fps is typical) minimize debris buildup and screen surface impingement.

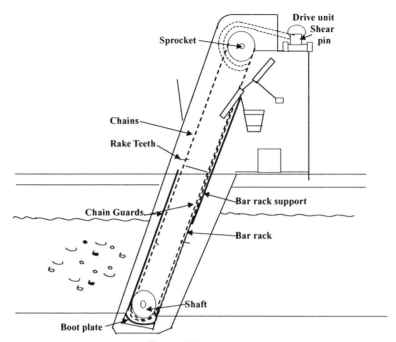

Figure 4.7 Bar screen.

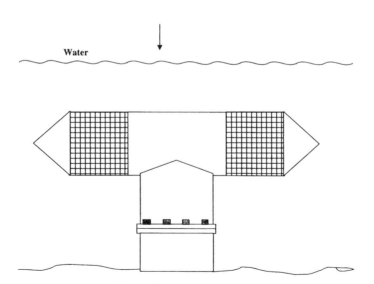

Figure 4.8 Passive intake screen.

Photo 4.7 Bar screen, lower left (Conestoga River Treatment Plant).

68

SUMMARY

Screening is a preliminary step that removes only relatively large visible pollution from the influent. The accumulation of trash that screening removes is less important in terms of sanitation of the water than the less apparent pollutants that remain. The downstream processes used to eliminate these problems are more complex, and eliminate far more dangerous problems from the raw water.

CHAPTER 5

Coagulation and Flocculation

5.1 PROCESS PURPOSE

COAGULATION is a chemical process that physically makes sedimentation more efficient (see Figure 5.1). Flocculation is a physical process used to enhance the effectiveness of coagulation's chemical addition. The screened influent is pumped into large *settling basins* (also called clarifiers or sedimentation tanks) where it is allowed to sit for a prescribed time. The settling basin allows gravity to handle many of the suspended impurities that remain in the influent. Under quiescent basin conditions, when flow and turbulence are at the minimum, *sedimentation* makes particles that are denser than water sink and settle to the tank bottom. Larger and heavier particles settle faster than smaller, lighter particles. The accumulation on the bottom of the tank is called sludge or water solids.

Not all suspended particles will be removed in this process, even under very long detention times. Buoyancy and drag (friction) affect the settling rate, as do water temperature and viscosity. Very small particles (including colloids, bacteria, color particles, and turbidity) will not settle out of suspension without some help. The coagulation process provides that help. By rapidly mixing coagulant chemicals with the water, then slowly and gently stirring the mixture before sedimentation, these particles form floc in the *flocculation* process. The larger, heavier floc particles settle, and can then be removed by subsequent settling and filtration. In fact, colloidal particles must be chemically coagulated to be removed (see Figure 5.2).

Coagulation and flocculation neutralize or reduce the natural repellent negative electrical charge that particles in water carry. This electrical charge keeps particles separate and in suspension. By chemically removing the charge, then

71

Lime hopper on plant roof.

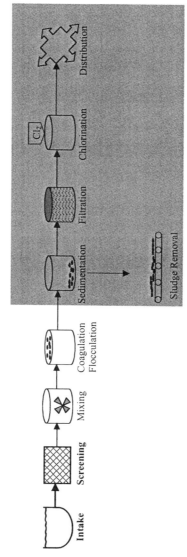

Figure 5.1 Basic water treatment unit processes: mixing and coagulation/flocculation.

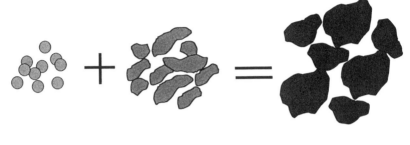

Flocculents, added to   mineral particles   create larger clumps,
which are easier to filter

Figure 5.2 "Clumping" of particles.

ensuring the particles contact, coagulation and flocculation alter the colloids so
that they adhere to form large floc particles.

5.2 COAGULANT CHEMICALS

Coagulation processes use coagulants and/or coagulant aids. The types appropriate for use are dependent on individual plant process schemes.

5.2.1 COAGULANT TYPES

Commonly used metal coagulants are based on aluminum (aluminum sulfate—alum—the most frequently used) (see Photo 5.1) or on iron (ferric sulfate). Table 5.1 lists other common coagulants.

Chemically, when alum is added to water, it ionizes, producing Al^{3+} ions, some of which neutralize the negative charges on the colloids. However, most of the aluminum ions react with alkalinity in water (bicarbonate) to form insoluble

TABLE 5.1. Common Coagulant Chemicals.

Common Name	Comments
Aluminum sulfate	Most common coagulant in the United States; often used with cationic polymers
Ferric chloride	May be more effective than alum in some applications
Ferric sulfate	Often used with lime softening
Ferrous sulfate	Less pH dependent than alum
Aluminum polymers	Synthetic polyelectrolytes; large molecules
Sodium aluminate	Used with alum to improve coagulation
Sodium silicate	Ingredient of activated silica coagulant aids

Source: Adapted from Larsen, 1990, p. 77.

Photo 5.1 Alum sulfate pumps, feeds, and piping.

75

aluminum hydroxide. The aluminum hydroxide absorbs ions from solution, and forms a precipitate and adsorbed sulfates.

AWWA's *Water Treatment: Principles and Practices of Water Supply Operations* (1995) describes the process step by step.

(1) Alum added to raw water reacts with the alkalinity naturally present to form jellylike floc particles of aluminum hydroxide, $Al(OH)_3$. A certain level of alkalinity is necessary for the reaction to occur. If not enough is naturally present, the alkalinity of the water must be increased.

(2) The positively charged trivalent aluminum ion neutralizes the negatively charged particles of color or turbidity. This occurs within 1 or 2 seconds after the chemical is added to the water, which is why rapid, thorough mixing is critical to good coagulation.

(3) Within a few seconds, the particles begin to attach to each other to form larger particles.

(4) The floc that is first formed consists of microfloc that still has a positive charge from the coagulant; the floc particles continue to neutralize negatively charged particles until they become neutral particles themselves.

(5) Finally, the microfloc particles begin to collide and stick together (agglomerate) to form larger, settleable floc particles. (p. 56)

5.2.2 COAGULANT AIDS

Coagulation and flocculation are somewhat delicate processes. When floc settles too slowly, or breaks apart too easily under the basin water movement, coagulant aids improve settling and floc toughness. The AWWA (1995) describes a coagulant aid as a chemical added during coagulation to improve coagulation; to build stronger, more settleable floc; to overcome the effect of temperature drops that slow coagulation; to reduce the amount of coagulant needed; and to reduce the amount of sludge produced.

Polymers (water-soluble, high-molecular weight organic compounds that carry multiple electrical charges along a chain of carbon atoms) are widely used coagulant aids that help build large floc prior to sedimentation and filtration. Activated silica, adsorbent-weighting agents, and oxidants are also used.

5.3 PROCESS OPERATION: COAGULATION

Coagulation chemicals are added in the rapid-mix tank, generally in a matter of minutes. Coagulant aids are added and blended into the electrically destabilized water during flocculation.

Water characteristics that affect this process include pH, temperature, and ionic strength. Chemical treatment is based on empirical data derived from jar testing or other laboratory tests and field studies (Viessman and Hammer, 1998).

5.4 PROCESS OPERATION: FLOCCULATION

Flocculation (the "clumping" of particles as the result of coagulation) is the most critical factor that affects particle removal efficiency. It speeds slow-settling particles and chemical precipitants.

This is accomplished by a slow mixing process, designed to bring particles into contact so that they collide, stick, and agglomerate (grow) to a size that readily settles. How fast and well the particles agglomerate depends on *velocity gradient*. Too much mixing shears the floc particles, breaking them down into smaller units. The velocity gradient (the speed the water is moved in the mixing process) must be strictly controlled. Enough mixing to bring the floc into contact must occur (the heavier the floc and the higher the suspended solids concentration, the more mixing is needed), without breaking apart the forming floc, until maximum floc formation occurs and sedimentation can begin. Gentle agitation for about one-half hour, using redwood paddles mounted horizontally on motor-driven shafts, is common. The paddles rotate slowly (about one revolution per minute), providing gentle agitation that encourages floc growth.

Floc formation depends on how much particulate matter is present, how much volume it occupies, and the basin velocity gradient. Experimentation with new techniques in coagulation and flocculation is on-going. A relatively new practice involves coagulation dispersion (flash mixing), flocculation, and sedimentation in a single unit called a *contact clarifier*.

SUMMARY

Coagulation and flocculation work together to remove much of the hidden debris that enters the plant in the influent. The smaller the impurities that remain, however, the more complex are the processes used to remove them.

REFERENCES

AWWA. *Water Treatment: Principles and Practices of Water Supply Operations*, 2nd ed. Denver: American Water Works Association, 1995.

Larsen, T. J. *Water Treatment Plant Design*. New York: Cox Publishing, 1990.

Viessman, W. and Hammer, M. J. *Water Supply and Pollution Control*, 6th ed. Menlo Park, CA: Addison-Wesley, 1998.

CHAPTER 6

Sedimentation

6.1 PROCESS PURPOSE

S EDIMENTATION (or clarification) is a physical process that separates settleable solids from influent by gravitational action. These solids include particulate matter, chemical floc, precipitates in suspension and other solids (see Figure 6.1).

6.2 PROCESS EQUIPMENT

Sedimentation takes place in settling or sedimentation tanks or basins, where water rises vertically for discharge through effluent channels in specific flow patterns suited to tank size and shape. In rectangular tanks, water flow will be rectilinear; center-fed and square settling tanks will operate with radial flow, and peripheral-feed basins use spiral flow. Design criteria include overflow rate, weir loading and detention time (1 to 10 hours is typical), and horizontal velocity for rectangular tanks. Equipment choice is determined using empirical data from full-size tank use performance. No matter what the equipment size and shape, settling tanks are designed for slow and steady water movement.

6.3 PROCESS OPERATION

Ideally, water flows horizontally through the sedimentation basin, then rises vertically to overflow the discharge channel weir at the tank surface (see Figure 6.2). While the water slowly rises, the floc settles in the opposite direction and is expelled mechanically by continuous-action sludge removal machinery. The

Water treatment sedimentation tank.

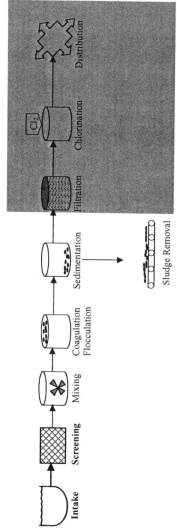

Figure 6.1 Basic water treatment unit processes: sedimentation added.

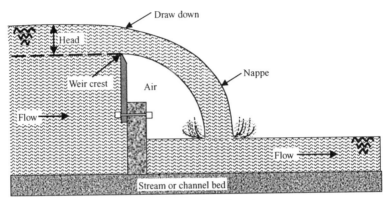

Figure 6.2 Side view of a weir.

process does not remove all of the floc. Particles lighter than the overflow rate flow out with the effluent, and are removed by filtration.

Sedimentation basin sludge is disposed of by passing it to a lagoon, thickening and holding, or directing it to the sanitary sewer (for more information on sludge and biosolids management, see Chapters 21 to 25).

SUMMARY

Sedimentation balances time efficiency against removal-rate efficiency. While the physical laws of gravity work in sedimentation, allowing the removal of materials too small for the initial screening to halt (through the combined efforts of coagulation and flocculation), the lighter materials are more efficiently removed by filtration in the next water treatment process stage.

CHAPTER 7

Filtration

7.1 PROCESS PURPOSE

FILTRATION is also a physical process, one that occurs naturally for groundwater sources. Surface waters percolate through porous layers of soil where they eventually recharge groundwater, reducing suspended matter and microorganisms to a level that ensures groundwater usually needs no treatment other than disinfection. Filtration of surface water is the last physical step in the process of producing potable water that meets the Safe Drinking Water Act turbidity requirement of 0.5 NTU (see Figure 7.1).

Generally, about 5% of the suspended solids and other impurities remain after sedimentation. This small percentage of remaining not-settleable floc particulate matter causes noticeable turbidity, and may shield microorganisms from disinfection. Filtration is a polishing process, and it involves passing the water through a layer (or bed) of porous granular material. This removes suspended particles by trapping them in the pore spaces of the filter media as the water flows through the filter bed (see Figure 7.2).

7.2 PROCESS EQUIPMENT

Filtration systems come in a variety of types and configurations, and system choice criteria include space considerations, speed considerations, and cost considerations for both installation and operation. Common filter types include rapid sand and slow sand filtration, diatomaceous earth filtration, and package filtration systems. Filters are also classified by the granular media (sand, anthracite coal, coal-sand, multi-layered, mixed-bed or diatomaceous earth)

83

Filter controls.

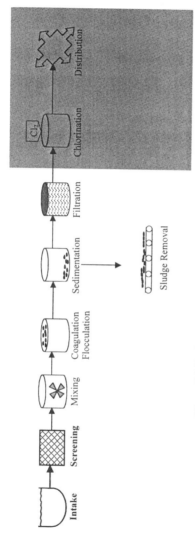

Figure 7.1 Basic water treatment unit processes: filtration added.

85

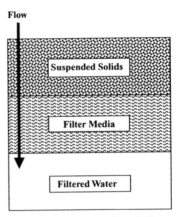

Figure 7.2 Water flow through filter.

and by directional flow (downflow, upflow, fine-to-coarse, or coarse-to-fine). Filtration action can work by either gravity or pressure.

Individual filter media present advantages and disadvantages. Ideally, the perfect filtration media qualities are that the media would have pore openings large enough to retain large quantities of floc to process large volumes of water without clogging, but small enough to prevent pass-through of suspended solids. The ideal filter media would also be light weight enough to allow sufficient depth for long filter runs, and be graded to allow effective backwash cleaning.

In gravity filtration, gravity forces the water down through the media, and forced upflow (backwashing) periodically removes collected impurities from the filter to clean it. Slow and rapid (refers to flow rate per surface area unit) sand filter systems are gravity filtration systems, and are commonly used for municipal applications.

In pressure filtration systems, the filters are completely enclosed to use line water pressure to push the influent through the filter.

7.2.1   SLOW SAND FILTRATION SYSTEMS

While slow sand filtration systems are reliable and use proven technology, modern plants generally don't employ them, because of the variety of problems associated with the systems. These problems are mostly related to small pore spaces in fine sand. While the small pores filter effectively, they also slow down the passage of water. Slow filtration time means increased filtration area is needed to process adequate amounts of water. This also means increased land usage to house the units. The fine pore spaces clog easily as well, requiring manual scraping to clean the filter (see Figure 7.3).

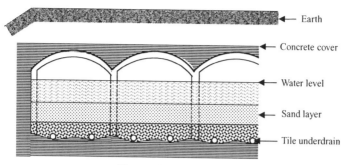

Figure 7.3 Slow sand filter.

7.2.2 RAPID SAND FILTRATION SYSTEMS

Rapid sand filters are the most commonly used systems for water supply treatment because of their reliability. They have replaced slow sand filters in most modern treatment plants. Rapid filters contain a layer of carefully sieved silica sand over a bed of graded gravel. The pore openings are often larger than the floc particles to be removed, so rapid filter systems use a combination of techniques to remove suspended solids and particulate matter from influent, including simple straining, adsorption, continued flocculation, and sedimentation. Filter cleaning is accomplished by daily backwashing (see Figure 7.4).

7.2.3 OTHER COMMON FILTRATION SYSTEMS

In *pressure filters*, as in rapid sand filters, water flows through granular media in a filter bed. However, pressure systems enclose the bed in a cylindrical steel tank and pump the water through the media under pressure. This can cause problems with reliability; occasionally solids are forced through the filter along with the effluent (see Figure 7.5).

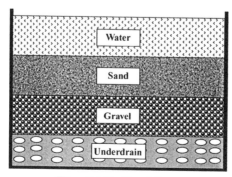

Figure 7.4 Rapid sand filter.

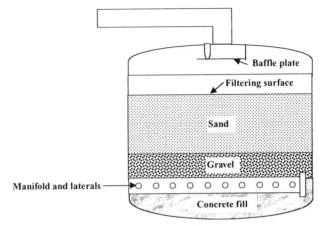

Figure 7.5 Pressure filter.

The *diatomaceous earth filter* contains a thin layer of a powderlike material formed from the shells of diatoms. These filtration systems also present reliability problems as well as expense considerations. Diatomaceous earth media is replaced rather than cleaned and reused by backwashing, raising operating costs.

Both pressure and diatomaceous earth (DE) filtration systems are more commonly used in industrial and swimming pool applications than they are in municipal installations.

SUMMARY

While filtration is an essential step in preparing the influent for final disinfection, it does not remove the most dangerous elements present in raw water. Bacteria and viruses are too minute to screen out. They will not coagulate, aggregate, settle out, or filter out without special processing. These pathogens must be deactivated by disinfection.

CHAPTER 8

Disinfection

8.1 PROCESS PURPOSE

DISINFECTION is a process (usually chemical) that deactivates virtually all recognized pathogenic microorganisms, but not necessarily all microbial life. The disinfection process accomplishes two things. *Primary* disinfection initially kills *Giardia* cysts, bacteria, and viruses. *Secondary* disinfection maintains a disinfectant residual that prevents regrowth of microorganisms in the water distribution system (see Figure 8.1).

Disinfection does not involve sterilization, which is the destruction of all microbial life. The level of disinfection needed for sterilization would be prohibitively expensive, and the end product would contain high levels of *disinfection by-products* (compounds formed by the reaction of a disinfectant such as chlorine with organic material in the water supply) and a strong chemical taste. Present treatment practices disinfect to the point that enough known disease-causing agents are eliminated to protect public health.

While chlorine disinfection is the best-known and most commonly used disinfection method, other methods are available. The three general types of disinfection include heat treatment (boiling water, commonly used in emergency situations), radiation treatment (UV radiation treatment), and chemical treatment (oxidizing agents including chlorine, ozone, bromine, iodine, and potassium permanganate, as well as metal ions such as silver, copper, mercury, and/or acids and alkalis).

Whatever disinfection method is used, the disinfectant chosen must possess certain characteristics. Obviously, it must work to kill off or deactivate pathogenic microorganisms. The disinfectant also

- must act in a reasonable time
- must act as temperature or pH changes

89

Sodium hypochlorite towers.

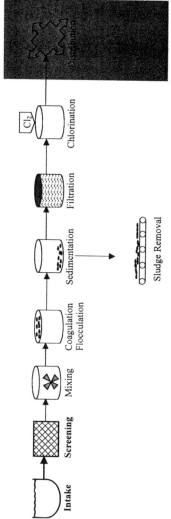

Figure 8.1 Basic water treatment unit processes: chlorination added.

91

- must be non-toxic
- must not add unpleasant taste or odor
- must be readily available
- must be safe and easy to handle and apply
- must be easy to determine the concentration of
- must provide residual protection
- must affect pathogenic organisms more than non-pathogens
- must be capable of continual application
- must disinfect with reasonable dosage levels to produce safe water

Disinfectants typically kill pathogens in one of five ways.

(1) The disinfectant damages the cell wall.

(2) The disinfectant alters the pathogen's ability to pass food and waste through the cell membrane.

(3) The disinfectant alters the cell protoplasm.

(4) The disinfectant inhibits cellular conversion of food to energy.

(5) The disinfectant inhibits reproduction.

8.2 CHLORINATION

In the United States, the disinfectant of choice has been chlorine. In general, chlorination is effective, relatively inexpensive, and provides effective levels of disinfectant residual for safe distribution. Applied as a gas (elemental chlorine, Cl_2), liquid (sodium hypochlorite), or solid (calcium hypochlorite), each of these forms has advantages and disadvantages.

The most cost-effective and efficient form (in terms of available chlorine) is *gaseous chlorine*. One volume of liquid chlorine under pressure will yield roughly 450 volumes of gas. Large treatment works commonly use this method, creating gas from liquid chlorine stored on site in high-pressure, high-strength steel cylinders (see Photo 8.1). Gaseous chlorination is also the most inherently dangerous method—chlorine gas is lethal at concentrations as low as 0.1% air by volume. In non-lethal concentrations, it irritates the eyes, nasal membranes, and the respiratory tract. Safety requirements for gaseous chlorine are extensive.

Because of the stiff safety requirements, and because it is easier to use and less toxic than gaseous chlorine, *sodium hypochlorite* (the form of chlorine in laundry bleach) is the most common disinfectant in smaller systems. Usually diluted with water before being applied as a disinfectant, sodium hypochlorite provides 5 to 15% available chlorine. Sodium hypochlorite must be handled

Photo 8.1 One ton chlorine tanks.

and stored with care, and its corrosiveness means it must be kept isolated from vulnerable machinery. Sodium hypochlorite solution costs more per pound of available chlorine, and provides lower levels of protection against pathogens than chlorine gas. Because of regulatory pressures, some large water treatment facilities are converting to sodium hypochlorite disinfection.

Calcium hypochlorite (a white solid available as tablets, powder, or in granular form) contains 65% available chlorine. Packaged calcium hypochlorite is stable, though it readily absorbs moisture from the air, reacting with it to form chlorine gas. It is also corrosive, strong-smelling, and requires proper handling. When in contact with organic materials (including wood, cloth, and petroleum products), chemical reactions can cause fire or explosion.

Though some forms of chlorine are safer to handle than others, using any form of chlorine for disinfection requires special care and skill from the operator.

Chlorine, whether elemental chlorine, sodium hypochlorite, or calcium hypochlorite, may be added to the incoming flow (prechlorination), or instead, added right before filtration. When used in prechlorination, chlorine works to help oxidize inorganics, and halts the biological action that occurs in the accumulations on the bottom of clarifiers, preventing dangerous gaseous build-ups. Chlorination prior to filtration keeps algae from growing and bacterial populations from developing in and on the filter itself.

8.2.1 CHLORINATION CHEMISTRY

In water, chlorine reacts with various substances or impurities present (organic materials, sulfides, ferrous iron, and nitrites, for example). The presence of these materials creates a *chlorine demand*, a measure of the amount of chlorine needed to eliminate these impurities by combining with them. That amount of chlorine cannot disinfect; it is already chemically depleted.

Chlorine also combines readily with ammonia or other nitrogen compounds, forming chlorine compounds. These chlorine compounds have some disinfectant properties, and are called the combined *available chlorine residual.* Chlorine in this form acts as a disinfectant. The chemically unchanged chlorine remaining in the water after combined residual is formed is *free available chlorine residual.* Free chlorine is much more effective than combined chlorine in disinfection.

For successful chlorination, several factors must be addressed: concentration of free chlorine, contact time, temperature, pH, and turbidity. How effectively chlorine disinfects is directly related to contact time with the water, as well as the free available chlorine concentration. Lower chlorine concentrations require increased contact times. Lower pH levels also aid disinfection effectiveness. Chlorine disinfects more quickly at higher temperatures. Turbidity affects chlorine's effectiveness as well, as it does any disinfectant. Chlorine must contact organisms to kill them. High turbidity levels provide shelter for microorganisms, preventing efficient contact.

8.2.2 CHLORINATION EQUIPMENT

Chlorine is usually fed continuously to the influent (see Photo 8.2). Probably the safest and most commonly used chlorine feed devices are all-vacuum chlorinators (see Figure 8.2). They are installed directly on the chlorine cylinder. The chlorinator ensures that gaseous chlorine under a partial vacuum in the line is carried to the point of application. Typically, the vacuum is formed by water flowing through the ejector unit at high velocity.

Usual application methods for hypochlorites involve adding the chemical in liquid form using positive-displacement pumps. These deliver a specific amount of liquid on each stroke of a piston or flexible diaphragm (for more information on chemical feeders, see Chapter 14, Section 14.5.2.1).

8.2.3 CHLORINATION BY-PRODUCTS

Although using chlorine for disinfection is not only common (an estimated 90% of U.S. water utilities use this method) and efficient, effective and relatively inexpensive, recent studies show risks related to the potential formation of chlorine by-products. Organic compounds (including decaying vegetation) combines with chlorine chemically, forming trihalomethanes (THMs) (see Figure 8.3). Chloroform (one of the THMs) is a suspected carcinogen, and other common trihalomethanes are chemically similar to chloroform, and raise concerns.

While many public utilities are exploring alternative methods of disinfection, others use approaches to reduce the possibility of chlorine by-product formation. Removing more of the organics before adding chlorine is a productive approach, as is simply changing the point in the treatment process where chlorine is added. These two approaches are generally accomplished by not chlorinating the raw water before filtration. Sometimes aeration or activated carbon adsorption are used to remove more organic materials. Reducing chlorine use to achieve a safe degree of disinfection with less chemical addition is also a possibility.

8.3 ALTERNATIVE METHODS OF DISINFECTION

The two most common alternative disinfection methods for water treatment are *UV radiation* and *ozonation*. Neither of these methods is an ideal solution for chlorine replacement, because of uncertainties and disadvantages with their use, including that they are not adequate for disinfection by themselves. While they do prevent the formation of THMs, both methods require secondary disinfection (usually chlorine) to maintain a residual during distribution.

Photo 8.2 Chlorinators.

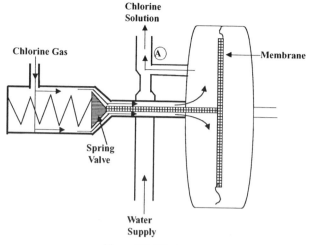

Figure 8.2 Chlorinator.

8.3.1 UV RADIATION

Ultraviolet or UV radiation (electromagnetic radiation beyond blue at the end of the light spectrum, outside the visible light range) is a physical process, not a chemical process—a big advantage over both chlorine and ozone as disinfectants. It disinfects by deactivating bacteria and viruses. The genetic material in microorganisms absorbs UV energy (UV light has a higher energy level than visible light), interfering with reproduction and survival. Turbidity severely affects UV radiation's ability to deactivate microorganisms, however, providing shelter for microorganisms from contact with the light.

Commonly, UV germicidal equipment consists of a series of submerged, low-pressure mercury lamps. Technological advances are making UV radiation a more viable disinfection alternative, both in terms of effectiveness and economics (see Figure 8.4).

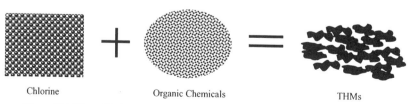

Chlorine   Organic Chemicals   THMs

Figure 8.3 When chlorine combines with organic compounds it forms trihalomethanes.

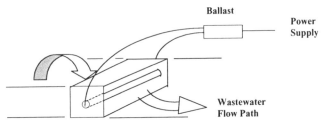

Figure 8.4 UV schematic.

8.3.2 OZONATION

Ozone (O_3) used as disinfectant leaves no taste and odor in the treated water, is actually more effective than chlorine against some viruses and cysts, and is unaffected by pH or ammonia levels in the water. Ozone is a gas at normal temperatures and pressures, and disinfects by breaking up molecules in water. When ozone reacts with organic materials and inorganic compounds in water, an oxygen (instead of a chlorine) atom is added, resulting in an environmentally acceptable compound. Ozone's instability, however, means that it cannot be stored and must be produced on site, generating higher costs

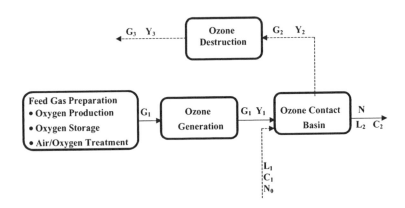

Legend:

G_1 Feed-Gas Flow Rate (m^3/min)
Y_1 Feed-Gas Ozone Concentration (g/m^3)
G_2 G_3 = Off-Gas and Exhaust-Gas Flow Rate (m^3/min)
Y_2 Off-Gas Ozone Concentration (g/m^3)
Y_3 Exhaust-Gas Ozone Concentration (g/m^3)
L_1 L_2 = Wastewater Flow Rate
C_1 Residual Ozone Concentration in Wastewater Influent (mg/L)
C_2 Residual Ozone Concentration in Wastewater Effluent (mg/L)
N_0 Influent Coliform Concentration (#/100 ml)
N Effluent Coliform Concentration (#/100 ml)

Figure 8.5 Simplified ozone process schematic diagram. *Source:* USEPA, (1986), p. 103.

than chlorine disinfection. Since ozonation also provides no disinfection residual, those equipment, labor, and chemical costs must also be factored in (see Figure 8.5).

8.4 MEMBRANE PROCESSES

In water treatment, membrane processes are generally used for demineralization. The two most common processes, reverse osmosis and electrodialysis, use microporous membranes to concentrate and separate the unwanted minerals from the influent.

Reverse osmosis and electrodialysis are related in concept to osmosis. In osmosis, a semipermeable membrane separates solutions of different mineral concentrations. Water will migrate through the membrane from the more dilute solution to the more concentrated solution until the hydrostatic pressure in the more concentrated solution is strong enough to stop the flow.

8.4.1 REVERSE OSMOSIS

In reverse osmosis (often abbreviated to R/O, and also called ultrafiltration), external pressure applied to the semipermeable membrane offsets the hydrostatic pressure, resulting in pure water on one side of the membrane and the unwanted mineral content concentrated on the other. While perhaps the most common use for reverse osmosis is to reduce salinity in brackish groundwaters, reverse osmosis is used to remove aesthetic contaminants that cause taste, odor, and color problems. It can also remove unwanted off-tastes caused by chlorides or sulfates, and is also used to treat for arsenic, asbestos, atrazine, fluoride, lead, mercury, nitrate, and radium. When paired with carbon prefiltering, reverse osmosis processes are used to remove volatile contaminants that include benzene, trichloroethylene, trihalomethanes, and radon.

Reverse osmosis equipment performance is affected by water quality parameters. Membranes are damaged by water constituents that include suspended solids, dissolved organics, hydrogen sulfide, iron, and strong oxidizing agents (chlorine, ozone, and permanganate) (see Figure 8.6).

8.4.2 ELECTRODIALYSIS

Electrodialysis demineralizes water using the principles of osmosis, separating anions and cations in solution by using ion-selective membranes and an electric field.

Both electrodialysis and reverse osmosis can be used to effectively treat water with high total dissolved solids concentrations, with appropriate pretreatment processes in place (see Figure 8.7).

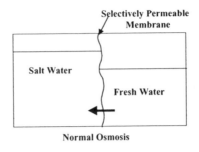

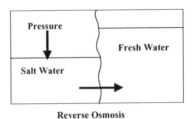

Figure 8.6 Normal osmosis and reverse osmosis.

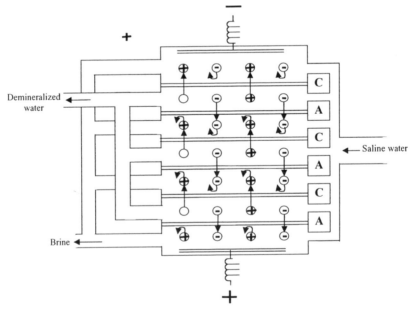

Figure 8.7 Schematic flow diagram of an electrodialysis system. Adapted from *Wastewater Treatment Plant, Planning, Design, and Operation*, 2nd ed., S. R. Qasim, 1999, p. 939.

SUMMARY

Once the water has been disinfected, it is ready for the consumer to use. Water travels from the treatment facility to the consumer's tap via the distribution system.

CHAPTER 9

Distribution

9.1 PROCESS PURPOSE AND METHOD

A municipality's water distribution and conveyance system serves two purposes: it carries raw water from a source to the plant, and it carries the finished water to the consumer (see Figure 9.1). Distribution systems consist of seven basic elements:

- sources (wells or surface water)
- storage facilities (reservoirs)
- transmission facilities for influent
- treatment facilities
- transmission facilities for effluent
- intermediate points (standpipes or water towers)
- distribution facilities

9.1.1 DISTRIBUTION SYSTEMS

Gravity distribution, pumping without storage, or pumping with storage are the three common distribution methods in use. When the water supply source is well above the community's elevation, *gravity distribution* is possible. The least desirable method, *pumping without storage* provides no reserve flow, and pressures fluctuate substantially. With this method, facilities must use sophisticated control systems to meet unpredictable demand. *Pumping with storage* is the most common method of distribution (McGhee, 1991).

In pumping with storage, water in typical community water supply systems is carried under pressure (pumping water up into tanks that store water at higher elevations than the households they serve provides water pressure) through a

103

Water tower in Chesapeake, Virginia.

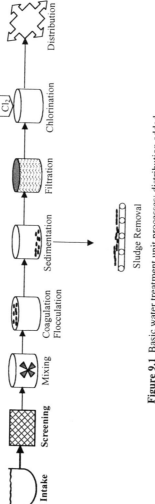

Figure 9.1 Basic water treatment unit processes: distribution added.

105

network of buried pipes. Street mains carry the water from standpipes or water towers to service individual business, industrial, commercial, or residential needs. Mains usually have a minimum diameter of 6–8 inches for adequate flows to supply buildings and for fire fighting. Pipes connected to buildings can be as small as 1 inch for small residences. House service lines (smaller pipes) from the main water lines transport water from the distribution network to households, where gravity's force moves the water into homes when household taps open. A primary goal for any water treatment facility is to provide enough water to meet system demands consistently, and at adequate pressures.

Distribution systems generally follow street patterns. The location of treatment facilities and storage works affects distribution, as do the types of residential, commercial, and industrial development present, and topography. Distribution systems commonly set up zones related to different ground elevations and service pressures. Water mains are generally designed in enclosed loops, to supply water to any point from at least two directions.

Distribution systems are categorized as grid systems, branching systems, or dead-end systems [see Figures 9.2(a)–(d)]. Grid systems are generally considered the best distribution system. The looped and interconnected arterials and secondary mains eliminate dead ends, and allow free water circulation so that a heavy discharge from one main allows drawing water from other pipes. Branching systems do not furnish supply to any point from at least two directions, and include several terminals or dead ends. In new distribution systems, antiquated dead-end systems are completely avoided. Older systems with terminals often incorporate proper looping during retrofitting.

9.2 PROCESS EQUIPMENT

Surface and groundwater water supply systems both generally involve canals, pipes, or other conveyances; pumping plants; distribution reservoirs or tanks to help balance water supply and demand, and to control pressures; other appurtenances; and treatment works.

Storage tanks for potable water distribution come in a variety of types (see Photos 9.1 and 9.2). Whatever the type, however, the tank interior must be properly protected and preserved from corrosion. Poor physical and material tank condition degrades the stored water. Any tank coating or preservative that will be in contact with potable water must meet the National Sanitation Foundation (NSF) Standard 61.

Tank types include:

(1) *Clear wells*—for storing filtered water from a treatment works. Also used as chlorine contact tanks (see Figure 9.3a).

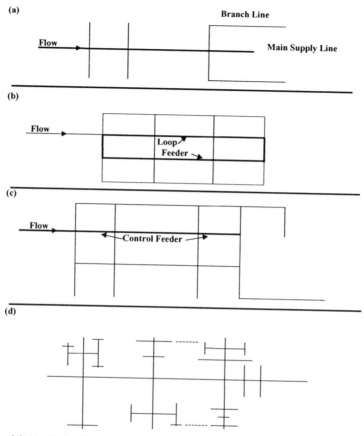

Figure 9.2 Distribution line networks. (a) Branched; (b) Grid; (c) combination branched/grid; (d) dead end.

(2) *Elevated tanks*—Primarily for maintaining an adequate and fairly uniform pressure to the service zone; elevated tanks are located above the service zone (see Figure 9.3b).

(3) *Stand pipes*—tanks that stand on the ground, with a height greater than their diameter (see Figure 9.3c).

(4) *Ground-level reservoirs*—maintain the required pressures when located above service area (see Figure 9.3d).

(5) *Hydropneumatic or pressure tanks*—often used in small water systems (with a well or booster pump) to maintain water pressures in the system and to control well pump or booster pump operation (see Figure 9.3e).

(6) *Surge tanks*—used mainly to control water hammer, or to regulate water flow (see Figure 9.3f), not necessarily as storage facilities.

Photo 9.1 Industrial water tower.

Photo 9.2 Water tower in Virginia Beach, Virginia.

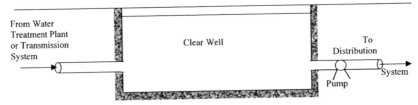

From Water
Treatment Plant
or Transmission
System

Clear Well

To
Distribution
System

Pump

Figure 9.3a Clear well.

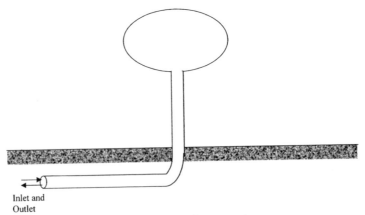

Inlet and
Outlet

Figure 9.3b Elevated storage tank.

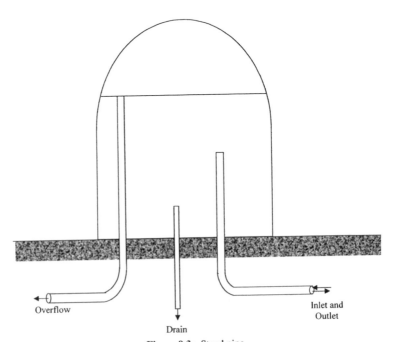

Overflow

Inlet and
Outlet

Drain

Figure 9.3c Stand pipe.

110

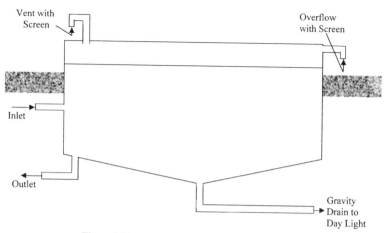

Figure 9.3d Ground-level service storage reservoir.

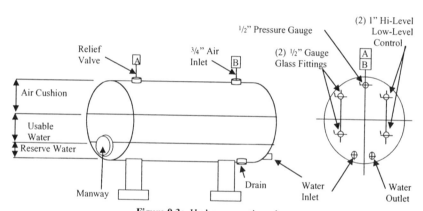

Figure 9.3e Hydropneumatic tank.

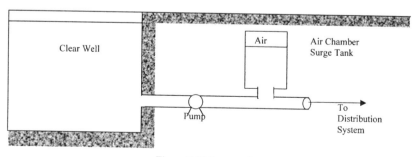

Figure 9.3f Surge tank.

Water stored in potable water storage facilities must be routinely properly monitored to detect problems in taste and odor, turbidity, color, and coliform presence. Monitoring includes determining chlorine residual levels, turbidity, color, coliform analysis, decimal-dilution, most probable number (MPN) analysis, and taste and odor analysis.

SUMMARY

Once a water treatment plant has delivered potable water to a consumer and the consumer begins to use it, whether that consumer is within a household, business, or industry, the water begins the journey that leads it to the other side of water treatment—wastewater treatment. Part III, "Basics of Wastewater Treatment," covers those treatment processes.

REFERENCE

McGhee, T. J. *Water Supply and Sewerage*, 6th ed. New York: McGraw-Hill, Inc., 1991.

BASICS OF WASTEWATER TREATMENT

Floating roof digester.

CHAPTER 10

Wastewater Regulations, Parameters, and Characteristics

10.1 PURPOSE: WASTEWATER PARAMETERS

THE outcome of the unit processes used to treat wastewater before outfall into a receiving body of water are controlled and determined by wastewater effluent quality parameters. Set by Congress through federal regulation, these parameters are supported and strengthened by state law. To maintain a permit to discharge into natural water systems, facilities must follow regulated programs of testing and reporting to prove their discharge will consistently meet regulatory standards.

10.2 PURPOSE: WASTEWATER TREATMENT

Wastewater treatment takes effluent from water users (consumers, whether from private homes, business, or industrial sources) as influent to wastewater treatment facilities. The wastestream is treated in a series of steps (unit processes, some similar to those used in treating raw water, and others that are more involved), then discharged (outfalled) to a receiving body, usually a river or stream.

Wastewater treatment takes the wastes and water that comprise the "wastestream" and restores the wastewater to its original quality. Wastewater treatment's goal is to treat the wastestream to the level that it is harmless to the receiving body. Most facilities actually set their goal higher: to treat the wastestream to achieve a water of a higher quality than the water contained in the receiving water body.

115

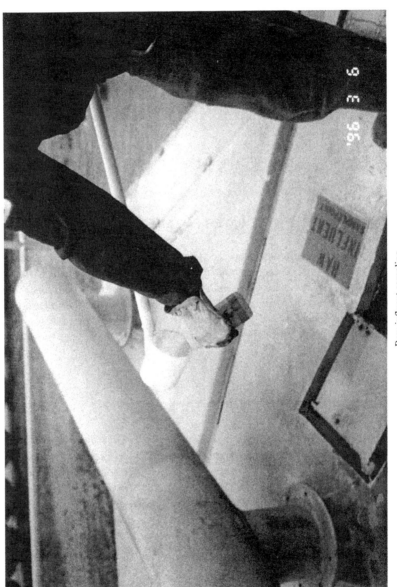

Raw influent sampling.

10.3 WASTEWATER REGULATIONS

The condition of the effluents discharged into water bodies, of course, affects the water supplies of communities downstream from the discharge point. The Water Pollution Control Amendments of 1972 radically changed how wastewater effluent was disposed of. The quality of effluent released into receiving waters significantly improved, positively affecting our water supply conditions.

With the Water Pollution Control Act of 1972 (Clean Water Act, CWA), the USEPA established standards for wastewater discharge. Municipal wastewater must be given secondary treatment, and most effluents must meet set conditions. Secondary treatment goals are set to the principal components of municipal wastewater, so that suspended solids, biodegradable material and pathogens can be reduced to acceptable levels. Industrial dischargers must use their industry's best available technology (BAT) to treat their wastewater.

The Clean Water Act also established a National Pollution Discharge Elimination System (NPDES) program based on uniform technological minimums for each point source discharger. The NPDES program issues discharge permits that reflect secondary treatment and best-available technology standards to each municipality and industry discharging effluent into streams.

10.4 WASTEWATER CHARACTERISTICS

Wastewater parameters provide a yardstick by which to assess the physical, chemical, and biological characteristics of wastewater. Meeting these parameters before discharge ensures the wastewater released into surface waters presents no chance of harm or disruption to the environment or to humans within a wide range of possible water uses.

10.4.1 PHYSICAL WASTEWATER CHARACTERISTICS

Wastewater physical characteristics of concern include the presence and quantity of solids in the wastestream, the degree of turbidity, the wastewater's color, temperature, and odor (see Figure 10.1).

10.4.1.1 Solids in Water

Solids removal is of great concern in wastewater treatment. Suspended materials provide adsorption sites for biological and chemical agents, and give microorganisms protection against chlorine disinfectants. As suspended solids degrade biologically, they can create objectionable by-products.

Solids can be either suspended or dissolved in water, and are classified by their size and state, by their chemical characteristics, and by their size distribution.

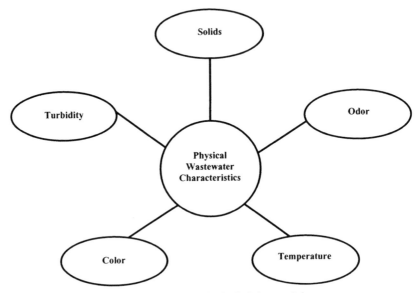

Figure 10.1 Wastewater's physical characteristics.

These solids consist of inorganic or organic particles, or of immiscible liquids such as oils and greases. Human water uses contribute other suspended materials. Large quantities of suspended solids (mostly organic in nature) are found in domestic wastewater. Industrial water use can add a wide variety of either organic or inorganic suspended impurities.

Filtration provides the most effective means of removing solids in treatment, although colloids and some other dissolved solids cannot be removed by filtration. Use of membrane technologies (reverse osmosis and electrodialysis) to remove dissolved solids is effective, and use of these processes is increasing.

The level of suspended solids (SS) is an important water quality parameter for wastewater treatment. It is used to monitor performance of several processes, to measure the quality of the wastewater influent, and to measure the quality of effluent. Most treated wastewater discharges must meet a maximum suspended-solids standard of 30 mg/L.

10.4.1.2 Turbidity

Water's clarity is usually measured against a turbidity index. Insoluble particulates scatter and absorb light rays, impeding the passage of light through water. Turbidity indexes measure light passage interference. Wastewater influent, of course, is expected to be turbid, but much of the materials that cause turbidity are removed with treatment.

Industrial and household wastewaters often contain many different turbidity-producing materials (detergents, soaps, and emulsifying agents); these are frequent wastewater constituents.

In wastewater treatment, whenever ultraviolet radiation (UV) is used in for disinfection, turbidity measurements are essential. UV light must penetrate the wastestream flow to effectively kill pathogenic microorganisms, and turbid wastestream flow reduces irradiation (penetration of light) effectiveness.

10.4.1.3 Color

Color is not a common concern for wastewater treatment, though it acts as wastewater condition indicator for judging wastewater's age. Ideally, wastewater early in the waste flow is a light brownish-gray color. The flow becomes increasingly more septic as travel time through the collection systems increases, and as more anaerobic conditions develop. As this occurs, wastewater's normal color changes from gray to dark gray and then to black.

10.4.1.4 Odor

Odor is a never-ending problem in wastewater treatment. Most urban treatment plants physically cover odor source areas (treatment basins, clarifiers, aeration basins, and contact tanks). This helps to prevent odors from leaving the unit processes, which helps to solve problems with local objections. However, these contained spaces can cause problems with toxic concentrations of gas. Such units must be vented to wet-chemical scrubbers to prevent toxic gas buildup.

10.4.1.5 Temperature

Temperature can play an important role in how efficiently wastewater unit treatment processes perform. Biological wastewater treatment systems are generally more efficient at higher temperatures. Temperature affects how quickly and effectively chemicals dissolve and chemical reaction times as well. In general, the colder the influent temperature, the more chemical is needed for treatment. However, summer heat increases chlorine demand, and promotes algae and microbial growth.

10.4.2 CHEMICAL WASTEWATER CHARACTERISTICS

The major chemical parameters of concern in wastewater treatment are total dissolved solids (TDS): alkalinity, metals, organics and nutrients, pH, and chlorides. These chemical parameters are directly related to the solvent capabilities of water (see Figure 10.2).

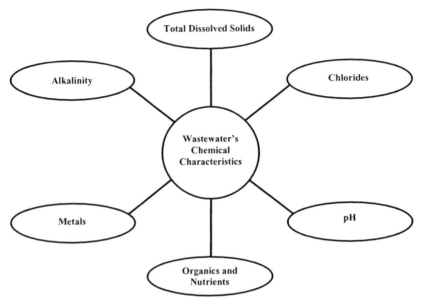

Figure 10.2 Waterwater's chemical characteristics.

10.4.2.1 Total Dissolved Solids

Solids in water occur either in solution or in suspension. The solids in the water that would remain after filtration and evaporation as residue are called total dissolved solids, or TDS. Dissolved solids can be removed from water by filtration and evaporation, and also by electrodialysis, reverse osmosis, or ion exchange. A routine solids test used in wastewater treatment to determine the efficiency of the treatment process is the measurement of settleable solids, the coarser fraction of the suspended solids that settle out from gravity.

10.4.2.2 Metals

Metals in wastewater (heavy metals such as cadmium, copper, lead, zinc, mercury, and others) are of high concern because they are often toxic to humans, and can be extremely harmful if discharged to the environment. They also can cause problems within treatment processes. Metals content is measured to determine toxicity levels.

Metals in sufficient concentrations will kill microorganisms in the activated sludge process. They are removed by chemical treatment, though that doesn't end the metals toxicity problem for wastestream processing. Industrial wastes with high levels of toxic metals or organic substances can contaminate sewage biosolids, thereby limiting biosolids disposal options and raising disposal costs.

10.4.2.3   Organics

The microbes that function in some biological treatment processes that rely on microbial decomposition consume dissolved oxygen (DO). This demand for oxygen is called the biochemical oxygen demand (BOD). BOD levels are measured by tests that determine how much DO aerobic decomposers use to decay organic materials in measured amounts of water held at 68°F (20°C) over a 5-day incubation period. Without continuous oxygen replacement, dissolved oxygen levels decrease until the cycle fails from lack of available oxygen as the microbes consume and decompose the organics.

"In wastewater of medium strength, about 75% of the suspended solids and 40% of the filterable solids are organic in nature," according to Metcalf & Eddy (1991, p. 65). Organics often present in wastewater include proteins, lipids (oils and grease), carbohydrates, and detergents. About 30+% of this organic matter is not biodegradable.

Proteins, which consist totally or in large part of many amino acids, also contain carbon, hydrogen, oxygen, sulfur, phosphorous, and a fairly high and constant proportion of nitrogen. The greater mass of wastewater biosolids material is made up of protein, or coated with protein so that chemically, it reacts as protein would. Most of wastewater's nitrogen comes from proteins and urea. Microorganisms decompose the nitrogen, in the process creating end products with foul odors.

Lipids (these include fats, oils, and waxes) are a heterogeneous collection of biochemical substances, soluble to varying degrees in organic solvents such as ether, ethanol, and acetone, but only marginally water soluble.

Food wastes contain high levels of lipids, mostly fats (compounds of alcohol and glycerol), oils, and grease. Fats are very stable organic compounds, and do not easily decompose.

Grease causes many problems in wastewater treatment. High amounts of grease can severely reduce the efficiency of filters, nozzles, and sand beds. Grease adheres to sedimentation tank walls, where it decomposes and adds to the amount of scum. When grease discharges with the effluent, surface water biological processes and aesthetics are affected.

Carbohydrates (widely found in nature, and a common component of wastewater) include starch, cellulose, sugars, and wood fibers. Some are soluble (sugars, for example), others are insoluble (starches, wood fibers). Lower organisms (including bacteria) use carbohydrates to synthesize fats and proteins, as well as for energy. Without oxygen, carbohydrate decomposition end products are organic acids—alcohols and gases that include carbon dioxide and hydrogen sulfide. Large quantities of organic acids hinder treatment by overburdening wastewater's buffering capacity, dropping pH levels, and halting biological activity.

Detergents (surfactants) are slightly soluble in water and cause foaming when out-falled in effluent into surface waters. Detergents can reduce the oxygen uptake in biological processes. Use of synthetic detergents have reduced or eliminated these problems.

10.4.2.4 Inorganics

Several inorganic constituents affect wastewater treatment. pH, chlorides, and nutrients including nitrogen and phosphorous, sulfur, toxic inorganic compounds, and heavy metals influence treatment processes.

10.4.2.4.1 pH

pH (hydrogen ion concentration) indicates the intensity of acidity or alkalinity in wastewater, and affects biological and chemical reactions. Wastewater's chemical balance (equilibrium relationships) is strongly influenced by pH. Weak acids, bases, and salts in wastewater cause wastewater's predictable behavior in treatment. For example, wastewater's pH levels directly impact certain unit processes, including disinfection with chlorine. Increased pH increases the contact time needed for chlorine disinfection.

10.4.2.4.2 Chlorides

Chloride (a major inorganic constituent in wastewater) generally does not cause any harmful effects on public health. Wastewater's chloride concentration is higher than raw water's from sodium chloride (salt), which commonly passes unchanged through the human digestive system.

10.4.2.4.3 Nutrients

The nutrients of greatest concern in wastewater treatment are nitrogen and phosphorous. Other nutrients include carbon, sulfur, calcium, iron, potassium, manganese, cobalt, and boron—all essential to the growth and reproduction of plants and animals.

As the primary component of the earth's atmosphere, nitrogen occurs in many forms in the environment and takes part in many biochemical reactions. Municipal wastewater discharges are a principle source of nitrogen in surface water, as is runoff from animal feedlots, fertilizer runoff, and some kinds of bacteria and blue-green algae that directly obtain atmospheric nitrogen.

Nitrogen in the form of nitrate (NO_3) in surface waters indicates contamination with sewage. An immediate health threat to both human and animal infants, excessive nitrate concentrations in drinking water can cause death.

Though the presence of phosphorous (P) released into drinking water supplies has little effect on human health, too much phosphorus in water supplies causes problems that contribute to algae bloom and lake eutrophication. Again, municipal wastewater discharges are a principle phosphorous source. Other sources include phosphates from detergents, fertilizers, and feedlot runoff.

10.4.3 BIOLOGICAL WASTEWATER CHARACTERISTICS

The presence or absence of pathogens in wastewater is of primary importance. Normal sewage contains millions of harmless microbes per milliliter. However, wastes from people infected with disease can produce harmful pathogenic organisms that then enter the sewage and the water systems receiving the sewage. While the processes that treat wastewater involve using the life-cycles of many types of bacteria and protozoa for removing the wastes from the water, the final effluent to be released must not carry dangerous levels of pathogens into the receiving waters (see Figure 10.3).

10.4.3.1 Bacteria

Bacteria are fundamental to several wastewater treatment unit processes, especially those responsible for degradation of organic matter. These include processes that occur in trickling filters, activated biosolids processes, and biosolids digestion. These bacteria must be controlled, however. Excessive growth of

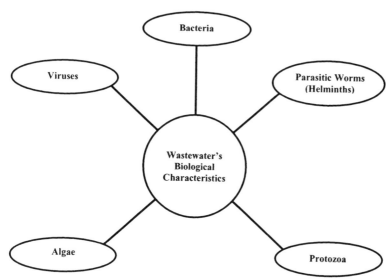

Figure 10.3 Biological water quality characteristics.

some species of ordinarily useful bacteria (for example, *Sphaerotilus natans*, a filamentous organism) can reduce treatment efficiency. Bacterial presence in effluent must also be controlled. Waterborne pathogenic bacteria transmit diseases that cause common symptoms of gastrointestinal disorder. Improperly treated effluent returns these pathogens to the water supply.

10.4.3.2 Viruses

Viruses cause special concern in wastewater treatment for several reasons. Turbidity in water provides shelter for pathogenic organisms from some disinfection methods. Testing for viruses in water is difficult because of limited identification methods—viruses are unpredictable in appearance and behavior. Viruses also present other problems. They are not easily trapped by standard filtration methods, because the many varieties of viruses are extremely minute.

Waterborne viral infections generally cause nervous system disorders, not gastrointestinal ones. Viruses must have a host to live and reproduce. They don't possess the machinery necessary for their own replication. They remain dormant until an appropriate "host" provides that machinery. Because of their ability to remain dormant until an appropriate host ingests them, viruses that reenter the water supply from wastewater effluent can cause problems for downstream water use.

10.4.3.3 Algae

Algae are found in wastewater, as well as in fresh water, saltwater and polluted water. Since most algae needs sunlight to live, they only grow near the water surface.

Algae play an important role in some kinds of wastewater treatment stabilization ponds. In aerobic and facultative ponds, algae (through photosynthesis) often supply the oxygen needed for microbial breakdown of wastes, and, in turn, use the waste products the bacteria and other microbial life leave behind as a food source—a natural, inexpensive, and simple way to remove wastes. However, algae growth is not as easy to control as aerators (the other common source of oxygen for aerobic or facultative treatment ponds). Algae rely on sunlight to produce oxygen and use up carbon dioxide. They use up oxygen and produce carbon dioxide at night and on dark days. In cold climates, algae die off in the winter. Heavy algae growths can also raise the level of suspended solids concentration in the effluent.

10.4.3.4 Protozoa

Protozoa are the simplest animal species, widely distributed and highly adaptable. Mobile, single-celled, completely self-contained organisms, some protozoa are free-living, while a few others are parasitic. Protozoan populations are an essential part of activated sludge treatment processes. Most of the protozoan population is removed by sedimentation after activated sludge treatment. While most protozoa are harmless—two important exceptions are *Entamoeba histolytica* (amebiasis), and *Giardia lamblia* (giardiasis)—effluent must not contain excessive protozoan levels. The cysts aquatic protozoans form during adverse environmental conditions make them difficult to deactivate by disinfection. Filtration is usually the most effective means to remove them from water.

10.4.3.5 Worms (Helminths)

Worms are organisms with aerobic requirements, that inhabit organic mud and slime. Frequent indicators of sewage contamination and pollution in streams, tubifex, bloodworms, and other helminths can metabolize solid organic matter that other microbes cannot degrade, and feed on sludge deposits. This helps to break down the organics in the waste stream. Parasitic worms are transmitted to humans (or to other carriers) through contact with untreated sewage or polluted waters.

10.4.3.6 Indicator Organisms

Testing and identification of pathogens present inherent problems. While some organisms are tough and persistent, and others form protective spores that allow them to resist treatment processes designed to kill off pathogens, most pathogens grow and multiply rapidly while within the human body, and do not survive long in nature. These qualities made finding a method for obtaining a clear indication of the presence or absence of pollution in a source water difficult. Individual pathogens may be present in such small quantities that they are undetectable. Much testing and retesting for each pathogenic organism would have to be done to determine safety. Testing for a single species solved the problem of determining the biological safety of water and wastewater. The indicator organisms (coliforms) alert us to the possible presence of sewage contamination. Coliforms present in a water sample indicate sewage contamination, and may also mean that the water source could contain pathogenic microorganisms that present a threat to public health.

10.4.3.7 Coliforms

Coliform group testing involves estimating the number of fecal coliform bacteria present in a measured water sample. This information provides a ratio of contamination relative to a defined quantity of water. The results of testing correlate to the amount of pollution in the source water—wastewater will contain more coliform groups than polluted water, which will contain more coliform groups than a water source intended for drinking water.

Fecal coliforms are present in the intestinal tract of all mammals, but do not themselves cause illness. Since fecal coliforms are always present in fecal wastes, water recently contaminated with sewage will always contain them. Coliforms always outnumber disease-producing organisms in contaminated water—human bodily wastes contain coliforms by the millions.

SUMMARY

Taking wastewater influent to accepted standards is the foundational task of wastewater treatment. The parameters and testing methods lay the groundwork for effective, efficient, wastewater treatment processes.

REFERENCE

Metcalf & Eddy. *Wastewater Engineering: Treatment, Disposal, Reuse*, 3rd ed. New York: McGraw-Hill, 1991.

CHAPTER 11

Wastewater Sources and Types

11.1 WASTEWATER

WASTEWATER is the flow of used water from a community, and includes household wastes, commercial and industrial wastestream flows, and stormwater and groundwater. By weight, wastewater is generally only about .06% solids—dissolved or suspended materials carried in the 99.94% water flow. This extreme ratio of water to solids is essential to transport solids though the collection system.

11.2 WASTEWATER SOURCES AND GENERAL CONSTITUENTS

The solids found in wastewater rarely contain only what most people consider sewage—human wastes. In fact, the dissolved and suspended solids sewage can contain varies widely from community to community, and is dependent, of course, on what inflow industrial and commercial facilities contribute to the treatment collection system. These influents mix with the more predictable residential flows, and provide so many possible substances and microorganisms to wastewater that complete identification of every wastewater constituent is not only rarely possible, but also rarely a necessary undertaking.

This is not to say that the solids an area's wastewater contains cannot be predicted in a general way. In most communities, wastewater enters the wastewater treatment system through one of five ways, each with its own usual characteristic solids loads. Common industry practice puts these constituents into several general categories.

127

Storm drain.

11.2.1 HUMAN AND ANIMAL WASTES

Generally thought the most dangerous wastewater constituent from a human health viewpoint, domestic wastewater contains the solid and liquid discharges of humans and animals. These contribute millions of bacteria, virus, and other organisms (some pathogenic) to the wastewater flow.

11.2.2 HOUSEHOLD WASTES

Domestic or residential wastewater flows also may contain paper, household cleaners, detergents, trash, garbage, and any other substance that a typical homeowner may pour or flush into the sewer system.

11.2.3 INDUSTRIAL WASTES

The materials that could be discharged from industrial processes into a collection system include chemicals, dyes, acids, alkalies, grit, detergents, and highly toxic materials. Individual industries present highly individual wastestreams, and these industry-specific characteristics depend on the industrial processes used. While many times industrial wastewaters can be treated within public treatment facilities without incident, often industries must provide some level of treatment prior to their wastestream entering a public treatment system. This prevents compliance problems for the treatment facility. An industry may also choose to provide pretreatment because their own on-site treatment is more economical than paying municipality fees for advanced treatment.

11.2.4 STORMWATER RUNOFF

In collection systems that carry both community wastes and stormwater runoff, during and after storms wastewater may contain large amounts of sand, gravel, road-salt, and other grit, as well as flood-levels of water (see Chapter 11 opening photo). Many communities install separate collection systems for stormwater runoff, in which case that influent should contain grit and street debris, but no domestic or sanitary wastes.

11.2.5 GROUNDWATER INFILTRATION

Old and improperly sealed collection systems may permit groundwater to enter the system through cracks, breaks or unsealed joints. This can add large amounts of water to the wastewater flow, as well as additional grit.

TABLE 11.1. Typical Composition of Untreated Domestic Wastewater.

Constituent	Abbreviation	Concentration (mg/L)
Biochemical oxygen demand	BOD_5	100–300
Chemical oxygen demand	COD	250–1000
Total dissolved solids	TDS	200–1000
Suspended solids	SS	100–350
Total Kjeldahl nitrogen	TKN	20–80
Total phosphorus (as P)	TP	5–20

Source: Adapted from Davis and Cornwell, 1991.

11.3 AVERAGE WASTEWATER PHYSICAL CHARACTERISTICS

The different specific substances that comprise wastewater vary in amount or concentration, depending on the source (see Table 11.1). However, an *average domestic wastewater* has the following physical and chemical characteristics.

11.3.1 PHYSICAL CHARACTERISTICS

- *Color:* typical wastewater is gray and cloudy. The color of septic wastewater changes to black.
- *Odor:* fresh domestic wastewater smells musty. Septic wastewater develops a rotten egg odor from the production of hydrogen sulfide.
- *Temperature:* normally, wastewater temperature remains close to that of the water supply. Infiltration or stormwater flow in significant amounts can cause major temperature changes.
- *Flow:* wastewater volume is normally expressed in "gallons per person per day," using an expected flow of 100 to 200 gallons per person per day. Expected flow rates are of principle concern in designing treatment plants. Expected flow figure may undergo revision to reflect the levels of infiltration or stormwater flow the facility receives. Flow rates can vary throughout the day as much as 50 to 200% of the average daily flow (*diurnal flow variation*).

11.3.2 CHEMICAL CHARACTERISTICS

- *Alkalinity:* an indicator of wastewater's capacity to neutralize acids, alkalinity is measured in terms of bicarbonate, carbonate, and hydroxide alkalinity. Alkalinity is essential to hold the neutral pH (buffer) of the wastewater during biological treatment.
- *Biochemical oxygen demand (BOD):* an indicator of the amount of biodegradable matter in the wastewater, normally BOD is measured in a 5-day test conducted at $20°C$ (BOD_5), and normally ranges from 100 to 300 mg/L.
- *Chemical oxygen demand (COD):* an indication of the amount of oxidizable matter present in the sample, the COD is normally in the range of 200 to 500 mg/L. Industrial wastes present in the wastewater can significantly increase this.

- *Dissolved gases:* the specific gases and normal concentrations dissolved in wastewater are based on wastewater composition, and under septic conditions may typically include oxygen in relatively low concentrations, carbon dioxide, and hydrogen sulfide.
- *Nitrogen compounds:* nitrogen's type and amount vary from raw wastewater to treated effluent, but nitrogen is mostly found in untreated wastewater in the forms of organic nitrogen and ammonia nitrogen (presence and levels determined by laboratory testing).
- *Total Kjeldahl nitrogen (TKN):* the sum of these two forms of nitrogen. Normal wastewater contains 20 to 85 mg/L of nitrogen, with organic nitrogen ranging from 8 to 35 mg/L and ammonia nitrogen ranging from 12 to 50 mg/L.
- *pH:* pH expresses wastewater's acid condition. For proper treatment, wastewater pH should generally range from 6.5 to 9.0.
- *Phosphorus:* in secondary treatment processes, phosphorus must be present in at least minimum quantities or the processes won't perform. However, excessive phosphorus causes stream damage and excessive algal growth. Phosphorus normally ranges from 6 to 20 mg/L. Removing phosphate compounds from detergents has significantly impacted the amounts of phosphorus found in wastewater.
- *Solids:* most wastewater pollutants can be classified as solids, and wastewater treatment is generally designed to either remove solids, or convert them to more stable or removable forms. General practice classifies solids by chemical composition as organic or inorganic, or by physical characteristics as settleable, floatable, or colloidal. Total solids concentration in wastewater normally ranges from 350 to 1,200 mg/L.
- *Water:* in even the strongest wastewater, the contamination present makes up less than 0.5% of the total. In wastewaters of average strength, contamination is normally less than 0.1%.

SUMMARY

Taking wastewater to accepted standards is the foundational task of wastewater treatment. Effective, efficient wastewater treatment removes the unwanted materials from inflowing waste streams, and prepares the wastewater for safe discharge into a receiving body of water. We examine an overview of the wastewater treatment process and the individual processes in the remaining chapters of Part 3.

REFERENCE

Davis, M. L. and Cornwell, D. A. *Introduction to Environmental Engineering.* New York: McGraw-Hill, 1991.

CHAPTER 12

Basic Overview of Wastewater Treatment

12.1 WASTEWATER TREATMENT

W ASTEWATER must be collected and conveyed to a treatment facility and treated to remove pollutants to a level of compliance with its NPDES permit before a municipal or industrial facility can discharge it into receiving water (see Figure 12.1).

The most common systems in wastewater treatment employ (as does water treatment) processes that combine physical, chemical, and biological methods. Wastewater treatment plants are usually classified as providing *primary, secondary*, or *tertiary* (or advanced) treatment, depending on the purification level to which they treat (see Table 12.1).

Once at a treatment facility, in primary treatment plants, physical processes (screening and sedimentation) remove a portion of the pollutants that settle or float. Pollutants too large to pass through simple screening devices are also removed, followed by disinfection. Primary treatment typically removes about 35% of the BOD and 60% of the suspended solids.

Secondary treatment plants use the physical processes employed by primary treatment, but augment the processes with the microbial oxidation of wastes. When properly operated, secondary treatment plants remove about 90% of the BOD and 90% of the suspended solids.

Advanced treatment processes are specialized, and their use is dependent upon the pollutants for removal. While usually advanced treatment follows primary and secondary treatment, in some cases (especially in industrial waste treatment), advanced treatment replaces conventional processes completely.

133

Wastewater treatment facility.

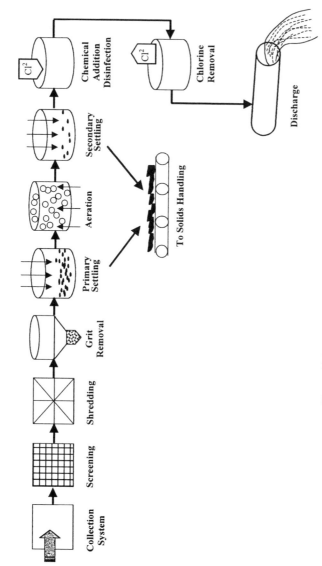

Figure 12.1 Unit processes for wastewater treatment.

135

TABLE 12.1. Wastewater Treatment Processes.

Process/Step	Purpose
Primary Treatment	Removes 90–95% settleable solids, 40 to 60% total suspended solids, and 25 to 35% BOD_5
Collection	Conveys wastewater from source to treatment plant
Screening	Removes debris that could foul or damage plant equipment
Shredding	Screening alternative that reduces solids to a size the plant equipment can handle
Grit removal	Removes gravel, sand, silt, and other gritty materials
Flow measurement	Provides compliance report data and treatment process information for hydraulic and organic loading calculations
Preaeration	Freshens septic wastes, reduces odors and corrosion, and improves solids separation and settling
Chemical addition	Reduces odors, neutralizes acids or bases, reduces corrosion, reduces BOD_5, improves solids and grease removal, reduces loading on the plant, and aids subsequent processes
Flow equalization	Reduces or removes the wide swings in flow rates for plant loadings
Primary sedimentation	Concentrates and removes settleable organic and floatable solids from wastewater
Secondary Treatment	Produces effluent with not more than 30 mg/L BOD_5 and 30 mg/L suspended solids
Biological treatment	Provides BOD removal beyond that achievable by primary treatment, using biological processes to convert dissolved, suspended and colloidal organic wastes to more stable solids
Secondary sedimentation	Removes the accumulated biomass that remains after secondary treatment
Tertiary or Advanced Treatment	Removes pollutants, including nitrogen, phosphorus, soluble COD, and heavy metals to meet discharge or reuse criteria with respect to specific parameters
Effluent polishing	Filtration or microstraining to remove additional BOD or TSS
Nitrogen removal	Removes nutrients to help control algal blooms in the receiving body
Phosphorus removal	Removes limiting nutrients that could affect the receiving body
Land application	Controlled land application used as an effective alternative to tertiary treatment methods. Reduces TSS, BOD, phosphorous and nitrogen compounds, as well as refractory organics
Disinfection	Destroys any pathogens in the effluent that survived treatment
Dechlorination	Protects aquatic life from high chlorine concentrations, needed to comply with various regulations
Discharge	Releases treated effluent back to the environment through evaporation, direct discharge, or beneficial reuse.
Solids treatment	Transforms sludge to biosolids for use as soil conditioners or amendments

136

SUMMARY

The following chapters describe the processes for wastewater collection, treatment, and wastewater reclamation and reuse, as well as those for biosolids management.

CHAPTER 13

Collection Systems[1]

13.1 PROCESS PURPOSE: COLLECTION

WASTEWATER collection systems carry wastewater (along with the solids accumulated in it) from the source (residential, commercial, or industrial) to the treatment facility for processing. Modern, fully enclosed sewage systems ensure that water contaminated with wastes and pollutants does not pose health, safety, or environmental problems (see Figure 13.1).

To handle the needs of a service area (the area that a sewerage system will service), to take advantage of gravity and the natural drainage afforded by area geography, and to lessen the costs of installing lift stations or pumps to move waste flows, treatment facilities are usually constructed in or near low-lying community outskirts, frequently along the edge of a natural waterway.

13.2 COLLECTION SYSTEM TYPES

Three types of sewerage systems are in general use: sanitary sewers, storm sewers, and combined sewer systems that carry both sanitary and stormwater flows.

13.2.1 SANITARY SEWERS

Sanitary sewers (by definition, sanitary sewers carry human wastes) convey wastewater from residences, businesses, and some industry to the treatment

[1]Information in this chapter is adapted from Parcher, Michael J. *Wastewater Collection System Maintenance*. Lancaster, PA: Technomic Publishing Co., Inc., 1998.

139

Sewer manhole.

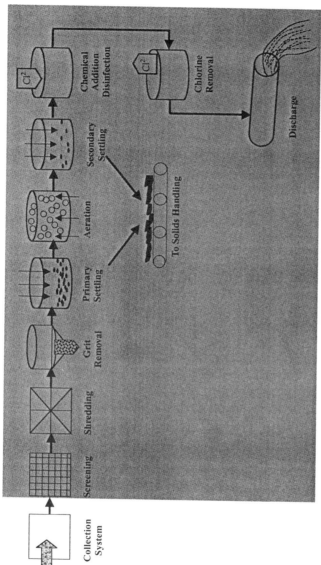

Figure 13.1 Unit processes for wastewater treatment: pretreatment.

141

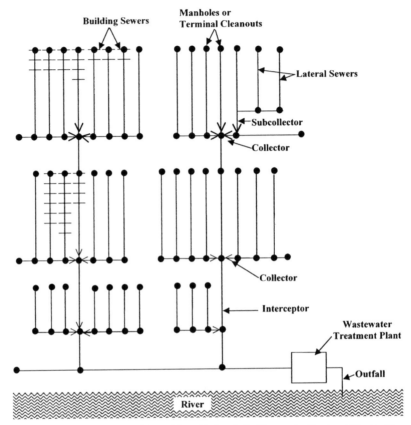

Figure 13.2 Collection system. Adapted from Qasim, S. R. *Wastewater Treatment Plants, Planning, Design, and Operation*, 2nd ed. Lancaster, PA: Technomic Publishing Co., Inc., 1999, p. 153.

facility. Unlike industrial waste flows, that may have some treatment prior to entry into a municipal system, the wastes sanitary sewerage carries are untreated. Of primary concern in sanitary sewerage management is preventing sewerage overflows, since these wastes contain infectious materials, and if released into the environment, cause serious risks to public health (see Figure 13.2).

13.2.2 STORM SEWERS

Storm sewers handle the influx of water into a collection system from surface runoff as the result of rainstorms or snowmelt. The more highly an area is developed, the more important effective stormwater collection systems are. As buildings and impermeable surfaces cover more area within a community, opportunities for storm flow to percolate into the ground to recharge groundwater are reduced, the heavier surface runoff becomes, and the more contaminants and pollutants can be carried by runoff. Storm systems must be designed to handle sudden heavy flows that can contain large quantities of sand, silt, grit, and gravel, as well as plant materials and trash (see Photos 13.1a and 13.1b).

As long as these flows do not carry infectious or human wastes, storm sewers can often be shunted, untreated, to natural drainage, although primary treatment may be required to meet NPDES permit requirements.

13.2.3 COMBINED SEWERAGE SYSTEMS

Combined sewerage systems carry both sanitary and stormwater flows. Combination systems always carry sanitary wastes, but are designed to handle large flows as well (commonly up to three times the average flows), so that during heavy rainfall, the same sewerage system can handle stormwater runoff as well as the normal sanitary flows.

While some older systems are still in operation, combined sewers are now seldom installed in the United States, because heavy precipitation can overwhelm the system, causing flows that exceed the treatment facility's ability to effectively treat them. Combined sewer overflows present a serious threat to public health.

13.3 COLLECTION SYSTEM COMPONENTS

A community's sewerage system consists of:

* building services that carry the wastes from the generation point to mains
* mains that carry the wastes to collection sewers
* collectors or subcollectors that carry the wastes to trunk lines
* trunk lines that carry the wastewater flow to interceptors
* interceptors that carry the wastes to the treatment plant
* other system elements that may include lift stations, manholes, vents, junction boxes, and cleanout points

All of these components except building connections are built under streets, easements, and right-of-ways, on layouts that take into consideration ground elevation, gradient, and natural drainage. They are designed to meet considerations of population size, estimated flow rates, minimum and maximum loads, velocity, slope, depth, and the need for additional system elements to ensure adequate system flows and access for maintenance.

Photo 13.1a Storm drain.

Photo 13.1b Storm drainage to the Chesapeake Bay.

145

- *Lift stations:* at points where gravity's force isn't enough to move wastewater through a system, lift stations are installed to pump the wastes to a higher point through a force main. Municipalities try to avoid installing lift stations; installation, operation, and maintenance of lift stations is expensive.
- *Manholes:* access into the sewerage system for inspection, preventive maintenance and repair is provided by manholes at regular intervals.
- *Vents:* gases that build up within sewer systems from the wastes they carry must be vented safely from the system. Human wastes in sanitary sewers carry sulfides, and hydrogen sulfide is deadly. Industrial wastes can carry risks related to the composition of their wastewater components.
- *Junction boxes:* sewerage systems are a network of piping, moving from small pipes that carry wastes from individual services, to larger mains, collectors, trunk lines, and interceptors. The constructions that occur when individual lines join are junction boxes or chambers. They are of special concern because leakage and infiltration can commonly occur at system joints (see Photo 13.2).
- *Cleanout points:* effective cleanout points provide access for cleaning equipment and maintenance into the sewer system.

13.4 CONSTRUCTION MATERIALS

Sewer lines are made from a wide variety of materials, and construction material selection is based on a variety of possible conditions and factors (see Photo 13.3).

13.4.1 MATERIALS

Rigid piping can be made from

- cast iron, ductile iron, corrugated steel, sheet steel
- concrete, reinforced concrete, asbestos-cement
- vitrified clay, brick masonry
- flexible piping
- plastics, including PVC, CPVC and other thermoset plastics, and polyolefins, polyethelene and other thermoplastics

Iron and steel piping offers the advantage of strength, but is affected by corrosion from both the wastes the pipes carry and from soil conditions. These materials are frequently used in exterior spans (piping runs that bridge gullies, for example). Concrete types offer high strength for heavy loading, especially for large volume pipes, but is heavy. Only short lengths of pipe are possible, so pipe runs must have many joints (joints provide areas in sewerage systems that allow for the advent of potential weak points through general deterioration, shifting, and root growth). Clay is highly resistant to corrosion, but is heavy and brittle. This type of piping is limited in length as well.

Photo 13.2 Junction boxes.

147

Photo 13.3 Interceptor pipes.

Plastics offer advantages of corrosion resistance, high strength to weight ratios, ease of handling, long pipe runs (so fewer joints), impermeability, and a certain amount of flexibility without loss of strength. They must be bedded carefully to avoid damage caused by soil voids.

13.4.2 SELECTION FACTORS

- how resistant the material is to corrosion
- how resistant the material is to flow and scour
- how resistant the material is to external and internal pressure
- soil conditions and backfill
- the potential wastewater load's chemical make-up
- requirements pertaining to strength, useful life, joint tightness, infiltration and inflow control, and other physical considerations
- costs, availability, ease of installation
- ease of maintenance

13.5 MAINTENANCE

Collection systems that are not properly and regularly maintained will cause interruptions in service in ways that will directly affect the residents in a community, and cause public outcry very quickly. This can be avoided with regular and thorough preventive maintenance.

Effective maintenance programs must accomplish two basic tasks. First, they should ensure that local residents, businesses, and industries (the collection system users) follow regulations and ordinances as to what may be properly and safely disposed of in the system, and how disposal is properly and safely accomplished, and that the collection system users meet plumbing codes and local ordinances that protect sewerage systems from damage and blockage. This is usually accomplished through a regular inspection program. The second part of an effective maintenance program involves a planned program of preventive maintenance and speedy repair. Preventive maintenance programs involve regular sewer line flushing and cleaning, clearing stoppages, and controlling the materials that cause odors or gases like hydrogen sulfide to build up, as well as maintaining the structural integrity of the system components.

13.6 LINE CLEANING

Well-designed sewer lines are self-cleaning. A combination of factors (including effective slope, velocity, and low friction coefficient) combine to ensure that the systems use the water in the wastes the systems carry to move the solids

TABLE 13.1. Line Cleaning Methods and Purposes.

Method	Purpose	Force Used	Water Pressure
Jetting	Moderate cleaning/ flushes light debris	Hydraulic/ mechanical	High pressure/ low volume
Flushing	Light cleaning/ poor debris removal	Hydraulic	Low pressure/ high volume
Balling	Light cleaning and debris removal	Hydraulic	Low pressure
Rodding	Clears heavy obstructions, debris removed other ways	Mechanical	None
Bucketing	Heavy debris removal	Mechanical	None

Adapted from Michael J. Parcher. *Wastewater Collection Systems Maintenance*. Lancaster, PA: Technomic Publishing Co., Inc., 1998, p. 46.

through the systems, rather than leaving the solids, sand, grit, grease, slime, roots, trash, and mineral build up behind to clog the lines.

However, not all sewers meet this ideal. Mechanical and hydraulic line cleaning are essential to maintaining the influent flow (Table 13.1). Mechanical cleaning moves tools through the lines to dislodge heavy debris. Hydraulic cleaning uses water pressure, flushing out lighter materials trapped within the pipes. All hydraulic cleaning tools runs the risk of flooding sewer services, and are only effective at removing light to medium accumulations of debris.

Line cleaning involves unblocking the line, then removing the debris that caused the line blockage in the first place. Light to moderate amounts of debris can be carried by water pressure to points in the collection system where they can be easily removed. Working through heavy debris or blockages takes mechanical cleaning methods; large amount of heavy debris usually must be removed by more labor-intensive means.

Collection system operators handle debris by flushing it into a trunk sewer, trapping it with screens, traps or "grit-catchers," vacuuming it out, manually hauling it out with bucket and shovel, pulling it out with gaffs or grabbers from topside, or using devices that winch the debris from the sewer line.

13.6.1 JETTING

Jetting is a versatile technique that effectively clears out moderate amounts of line blockage. Jetting cleans and flushes the line in a single operation, using a high pressure hose and a variety of nozzles to combine the advantages of hydraulic cleaning with mechanical cleaning.

Jetting is useful for routine line cleaning, for scheduled cleaning of high-problem lines, for quickly clearing sewer backups, and for clearing sewer lines after construction or roadwork.

Basic jetting equipment includes a jet truck that carries a water tank, a high pressure-high-volume pump, around 600 feet of 1-inch hose, and a variety of nozzles. In operation, the hose and nozzle are set in the manhole, facing upstream. When the pump is started, it drives the nozzle through the line to the next manhole. At that point, the hose is pulled back, still pushing water through the nozzle and carrying the debris within the line back to the first manhole for removal.

Different types of nozzles extend jetting capabilities. When lines are regularly maintained and obstructions are not extensive, jetting with the proper nozzle type can effectively clear a variety of obstructions. Hydraulic root saws and other cutting tools, pipe brushes, chain knockers, tornado spinner nozzles, skids, cages, nozzle extensions, and bangers are mechanical aids used with jetters to clear line blockages.

13.6.2 FLUSHING

Flushing adds large volumes of water to the sewer at low pressures. While useful after rodding for pushing leftover loose debris downstream, flushing uses water inefficiently, and does not effectively move accumulated solids. Jetting is generally a better use of resources for regular line cleaning.

In tank-dumping, the contents of a jet truck tank or water tanker are dumped through a manhole to clear a single run. Any effective water pressure is quickly lost as the water moves downstream.

Power flushing (the most efficient flushing technique) uses an open-ended jet hose to add water, accelerate wastewater movement, and flush grit into a downstream main. Power flushing can effectively clean outfall lines, and is useful for cleaning in hard-to-access sewers.

Flow from a hydrant can also be directed into a manhole for line-flushing, temporarily increasing water volume. With hydrant use, care must be taken not to make an illegal cross-connection between the potable water system and the collection system by laying the firehose line into the manhole.

13.6.3 BALLING

Balls (and other hydraulic line-cleaning devices, including kites, pills, pigs, scooters, and bags) work by partially plugging a flooded upstream main. Attached to a cable or rope to control its downstream path, the ball (or other cleaning device) moves through the pipe by upstream water pressure. As the water rushes past the cleaning device, movement of both the tool and the partially trapped water loosen debris, and the water flushes it downstream, where it can be collected and removed at the downstream manhole.

In general, balls and pills are most effective at clearing soft solids and light sludge cleaning in small diameter mains. Kites and bags (parachute-like devices

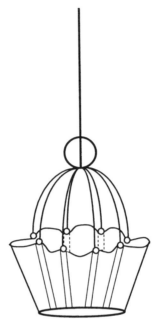

Figure 13.3 Debris catcher basket. Adapted from Parcher, M. *Wastewater Collection Systems Maintenance.* Lancaster, PA: Technomic Publishing Co., Inc., 1998, p. 248.

that open within the line to partially block the sewer) work more effectively than pills or balls for large-diameter lines (see Figure 13.3). Scooters and tires are heavier devices that operate under the same mechanical principles—the obstruction holds back an increased water supply, allowing the water pressure head to both clear accumulated obstructions and move the device down the line, while the operators control the device movement rate by rope, cable, or chains. Pigs, used for cleaning water mains and pressure sewers, are not restrained, but are inserted into the pipe, driven through by adding hydrant flow behind the device, then retrieved at the end of the pipe.

13.6.4 RODDING

Rodders can clear obstructions like heavy root accumulations that jetting can't handle. They work well to remove large soft obstructions to restore flow. Brush attachments can also be attached to rods to scour the pipe after clearing roots or other obstructions. Rodders don't need water to remove debris, and can work either with water flow or against it.

Rodding uses two basic movements to work through an obstruction, rotation (torque) and pushing against and pulling back on the obstruction (thrust). Mechanical rod drivers push the rod into the line from the surface, while pinch rollers mechanically thrust the line in and out of the pipe. Rods can be sectional or continuous. Rodding equipment can be either truck or trailer mounted and is powered by either the truck engine or an auxiliary engine.

13.6.5 BUCKETING

To clean large diameter mains of heavy solids accumulation, operators can winch a bucket through the line. This technique is simple, and the most effective method for cleaning large amounts of debris from a line, though it is labor-intensive and dangerous because of confined spaces concerns. While operators enter the line with the bucket attached to a line and shovel the solids accumulation onto the bucket, top operators pull out the full bucket and dump it, returning it to the line for another load.

13.7 COMMON COLLECTION SYSTEM PROBLEMS

All collection systems have to deal with the same problems, although some communities may have their normal, everyday sewerage problems compounded by problems caused by aging infrastructures. Some sewerage problems are caused by the nature of the materials collection systems carry, and others by the environment in which collection systems must be built.

13.7.1 AGING SYSTEMS

Many sewerage systems in operation today were constructed in the 1950s and 1960s (see Photos 13.4a and 13.4b). Keeping these aging systems in operation can present a challenge; overcoming the decay of the original materials is expensive whether it is approached through retrofitting or replacement, or simply by keeping after the maintenance. Loosened joints, cracked tiles, and corroded concrete let in root growth and inflow and infiltration. Old sewerage lines are frequently less efficient, trapping solids and grease and clogging more easily. Population growth may mean a system has outgrown its ability to effectively handle the current population.

13.7.2 ROOTS

Wastewater collections systems, installed for the most part underground, share their underground environment with various elements that are not always physically compatible with system materials or construction. Shifting earth, water infiltration from the surface or from groundwater all cause problems, but these problems are mostly passive. Of prevalent concern in sewerage systems are the problems caused by the root systems of natural vegetation—in short, trees.

Photo 13.4a Removing broken interceptor pipe.

Photo 13.4b Replacing interceptor pipe.

Tree roots mindlessly and insidiously grow into the soil, working rootlets into the smallest cracks and crevices, and then expanding the cracks as they grow larger. They are also attracted to water. An aging joint in a sewer pipe, or a joint that is not tight is an open invitation to root invasion, and the invading roots can grow to monster proportions relatively quickly.

Root growth in collection systems must be controlled. Regular line cleaning can clear root growth before it builds to a serious blockage, and frequent inspection and cleaning of noted problem areas helps keep known problems under control.

Techniques for removing roots include jetting lines with root cutters for light to moderate root growth, rodding to remove heavy root growth, and chemical treatments to control re-growth.

13.7.3 GREASE

Oil and water don't mix—in the sewer line or anywhere else. Grease, oils, and fats are a common and never-ending problem for collection systems, and in wastewater treatment in general. Grease sources are residential (usually light, expected, and manageable), commercial, and industrial. Restaurants as well as businesses that are sources of petroleum fuels, oils, and grease need special arrangements to control the wastes that enter the collection system.

Grease traps, sand and oil traps, and interceptors are used to prevent these wastes from entering the collection system and causing problems downstream.

13.8 NEW TECHNOLOGIES

New equipment and technology is rapidly changing the way that collection systems are inspected, maintained, and repaired. Electronic tools have increased visual access to the sewer lines, and have made possible increased use of more accurate mapping and database tools. Trenchless technologies are changing the methods by which sewerage systems are repaired or upgraded.

13.8.1 ELECTRONIC TECHNOLOGY FOR COLLECTION SYSTEMS

The task of visually inspecting sewer lines presents some challenges. While sewer lamping, manhole inspection, and large sewer-main entries are still used, technologies of miniaturization have provided collection systems with new tools to inspect and evaluate sewer lines, including visual entry into piping too small to effectively view in the past.

Specialized video equipment allows operators to assess line conditions quickly and accurately. With a waterproof video camera, lights, a system to transport the camera through the lines, and a closed-circuit TV system, sewer line conditions can be viewed, evaluated, and recorded. With sophisticated camera control equipment and robotics, operators can finally get a good look at problems far out of visual reach.

Video technology is not the only modern technology changing the way sewerage systems are managed. Computer record keeping, maintenance programs and databases allow easy access to more information than was previously possible. Computer mapping techniques ensure that the information operators take out on the job is current and accurate. Geographic information systems (GIS) provide spatial referencing that was not feasible only a few years ago.

13.8.2 TRENCHLESS TECHNOLOGIES

The costs, dangers, and disruptions that traditional open cut work causes when installing new lines, making repairs, or performing line rehabilitation make new trenchless technologies a very attractive alternative. These techniques are now more or less competitive in price, and new techniques, materials, equipment, and tools are increasing the useful possibilities for sewerage repairs and upgrades.

While individual methods present advantages and disadvantages, such techniques as horizontal directional drilling (HDD) and cured in place liners are offering alternatives to digging out lines for traditional excavation and open trench pipe laying or replacement.

For new installations, HDD, pipe jacking, tunneling, microtunneling, and auger boring in general drill or bore a tunnel through the soil and pull or push new pipe in behind the drill head. In a similar technology for replacing old pipe, pipe bursting destroys the old pipe while pulling the new pipe in behind.

To repair or replace existing lines with trenchless technology, all operations are carried out through existing manhole entrances. Cured-in-place pipe, man-entry liners, spiral wound liner strips, and segmental liners are trenchless techniques now useful for replacement or repair. Thermosetting resins, delivered to the needed repair location by a wide variety of methods, are used to reline existing pipes, and inversion and winched-in-place methods are used to install cured-in-place pipes. Tap cutting robots (remote controlled, air powered routers) are used to clear the new liner from the taps.

Sliplining slides a new, smaller diameter, polyethelene pipe liner into old damaged pipe. Because of the long pipe runs and butt fusion joints possible with polyethelene, this method is very popular.

The new electronic methods of line evaluation have made exacting point repair possible. While in the past, exactly which segment of a stretch of pipe was damaged was often a matter of guesswork. Now, with video technology and GIS mapping, the exact segment can be determined, and repairs effected that disturb only that point, rather than rehabilitating a whole run. Often the exact trouble spot can be pinpointed and accessed directly, rather than the involved and expensive process of digging up long stretches of piping to locate and repair the source of the problem.

SUMMARY

Wastewater collection starts wastewater's journey from potable water use and discard to outfall. Once wastewater is collected, the processes necessary for treatment can occur. Pretreatment of the influent (the subject of Chapter 14) begins the active treatment processes.

REFERENCE

Parcher, M. J. *Wastewater Collection Systems Maintenance.* Lancaster, PA: Technomic Publishing Co., Inc., 1998.

Preliminary Treatment

14.1 PRELIMINARY TREATMENT PROCESSES

PRELIMINARY treatment may include several processes, each designed to remove specific materials. Selection of what processes are included in pretreatment depends on the collection system (e.g., combined sewer systems, which collect stormwater, bring more grit into the influent than do sanitary sewers), and upon the nature or make-up of the wastewater in the community. Moreover, the balance between residential wastewater contents and industrial wastewater contents, and the individual industries that add wastes to the system will be determining factors in process selection for a treatment system.

Once the influent has entered the treatment facility via the collection system, pretreatment may include some or all of the following processes: screening, shredding, grit removal, flow measurement, preaeration, chemical addition, and flow equalization (see Figure 14.1).

14.2 PROCESS PURPOSE AND EQUIPMENT

Preliminary treatment provides the first rough pass at removing waste solids from the waste stream. Each pretreatment unit process works to remove a specific kind of material that should be eliminated before the influent flows on to downstream unit processes.

14.2.1 SCREENING

Coarse screening is the first unit process in wastewater treatment, as it is in water treatment, for the same reasons (see figures in Chapter 4). Screening

159

Bar screen.

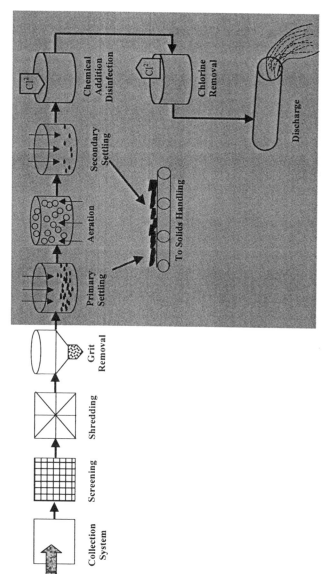

Figure 14.1 Unit processes for wastewater treatment: pretreatment.

161

removes large solids from the flow. These may include natural and man-made trash (leaves, branches, roots, rocks, rags, and cans). From each million gallon of influent, a typical plant could remove from 0.5 to 12 ft^3 of screenings.

Wastewater screening generally employs a *bar screen*, which consists of parallel evenly spaced metal bars, or a perforated screen. Bar screens can be coarse (2- to 4-inch spacing) or fine (0.75- to 2.0-inch spacing). Placed across a channel, the wastestream carries through the screen, leaving behind trapped large solids for removal. Solids removal from the screens may be either manual or mechanical, but should occur frequently enough that trash build-up does not block influent flow. Screens that are cleaned manually are placed at a 30° angle for ease of cleaning; those mechanically cleaned are placed at a 45 to 60° angle for improved mechanical operation (see Photo 14.1).

Plant design, solids load, and whether or not screening should be intermittent, constant, or only for emergency use are factors in determining what screening methods are used for a facility.

14.2.1.1 Shredding

Shredding approaches the problems that large solids create differently than does screening. Shredding reduces solids to a size that can enter the plant without damage to equipment or process interruption. The two common shredding processes used in wastewater treatment are comminution and barminution.

Wastewater treatment facilities generally prefer comminution devices for shredding. The entire influent stream flows through the comminuter's grinder assembly (objects too large to fit through the entry slots or that float are shunted aside and must be removed manually). Grinder assemblies include a screen or slotted basket, and two cutters, one rotating or oscillating and the second stationary. The solids are shredded between the cutters and pass through the screen or slots for removal in downstream processes.

Barminution combines bar screening with comminution. The solids collected on the bar screen are shredded, then passed along to downstream processes.

In operation of both comminution and barminution devices, proper cutter alignment and edge are essential factors in effective operation. Checking and maintaining alignment, and cutter sharpening or replacement are common maintenance needs for these devices.

14.2.2 GRIT REMOVAL

Wastewater influent (especially influent from combined sewer systems) may carry gritty materials (sand, silt, coffee grounds, eggshells, and other inert materials). Heavier than organic solids, these solids may cause excessive equipment wear (e.g., pump impellers). Grit removal takes these materials out of the wastestream.

Photo 14.1 Bar screen.

All grit removal methods rely on the grit's weight. Since grit is heavier than organic solids, the organic solids can be kept in suspension to be carried on for treatment, while the grit is separated out by processes that employ gravity/velocity, aeration, or centrifugal force. Some grit removal processes take place in a channel or tank, others in a centrifugal chamber.

Gravity/velocity controlled grit removal takes place in a tank or channel. Within the enclosure, the velocity of the wastewater is strictly controlled, to an ideal rate of about one foot per second. As long as the wastewater velocity remains between 0.7 and 1.4 feet per second (fps), the grit will settle and the organic material stay suspended. Velocity control is maintained by the amount of water flowing through the channel, by channel width and depth, or by the cumulative width of service channels (see Photo 14.2).

Cleaning of gravity systems can be manual or mechanical. Generally, manual cleaning involves ventilating the channel, taking the channel out of service and draining it, then cleaning. Mechanical cleaning is either continuous or on a timed cycle. Frequency depends on grit build-up, and cleaning should occur often enough to prevent grit carry-over into downstream processes.

Aeration keeps inorganic suspended solids in suspension in *aerated grit removal systems*, allowing heavier grit particles to settle out.

The aeration rate is determined by observing mixing and aeration, and sampling fixed suspended solids. In practice, the aeration rate is adjusted to produce the desired separation. Too much aeration keeps both the grit and organic materials in suspension. Too little aeration allows both the grit and the organics to settle out.

While aerated grit removal systems can be manually cleaned, the majority of these systems are mechanically cleaned.

Cyclone degritters separate heavier grit particles from lighter organic particles by centrifugal force. Usually used on primary sludge rather than the entire wastewater stream, the critical process control factor is inlet pressure. Excessive pressure will flood the unit and carry grit with the flow. Separated grit discharges directly into a storage container.

14.2.3   FLOW MEASUREMENT

Wastewater treatment processes use flow measurement to ensure process efficiency, to provide information for hydraulic and organic loading, and for data needed to prepare regulatory compliance reports. Flow rates are measured by finding a physical measurement that can be related to the quantity of liquid moving past a given point in a specified length of time. The three most common methods are fill and draw, weirs, and flumes.

The *fill and draw* method measures the time required for liquid to fill a container of known volume. Dividing the liquid volume by the required time provides the flow rate.

Photo 14.2 Screw pump for grit.

165

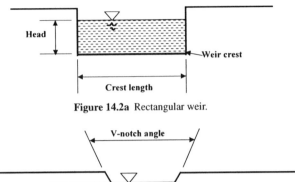

Figure 14.2a Rectangular weir.

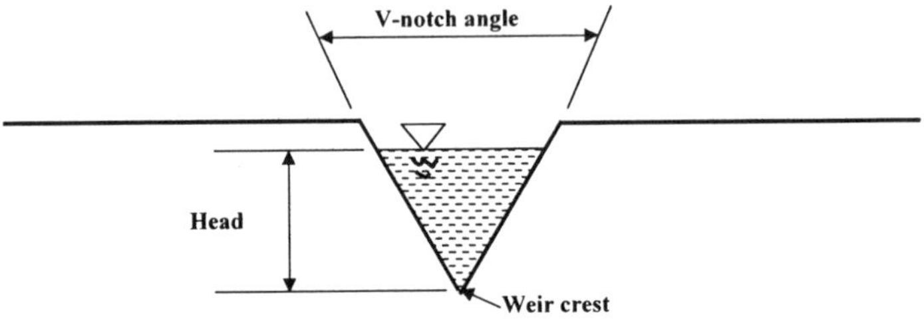

Figure 14.2b Triangular v-notch weir.

Weirs and *flumes* both use the principles involved in placing a constriction or barrier in an open channel. The amount of water that passes over or through the constriction is directly proportional to the height of the water behind the constriction (the head) and the area of the flow opening in the constriction. The constriction's opening area remains constant, so the only required measurement for calculating flow rate is the head behind the constriction (see Figures 14.2a and 14.2b).

In flume design, the throat of the flume (a narrow section) widens gradually in the converging section to the width of the channel. The flume's throat produces a head in the converging section that is measured to convert to the flow rate. Critical depth is measured at the flume. For *Parshall flumes* (the most common flume type), the head is measured at about two-thirds the length of the converging section, either manually or mechanically, and flow rates can be read from a chart, calculated or determined mechanically or electronically (see Figure 14.3).

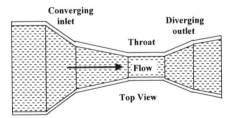

Figure 14.3 Top view of a Parshall flume.

14.2.4 PREAERATION

The process of preaeration forces air through the wastestream, achieving and maintaining an aerobic state. Aeration for 10 to 30 minutes freshens septic wastes, reduces odors and corrosion by stripping off hydrogen sulfide, and improves solids separation and settling by agitating the solids to release trapped gases. Preaeration for 45 to 60 minutes also reduces BOD_5. Aeration tank or channel blowers send air through diffusers on the tank bottom. The air bubbles carry trapped gases and hydrogen sulfide with them as they travel to the wastestream surface. Ideal air bubble size and rate depend on individual process and wastestream needs, but in general, the smaller the air bubbles, the more air introduced into the tank, and the more effective the process.

14.2.5 CHEMICAL ADDITION

Chemicals used in primary treatment include peroxide, acids and bases, mineral salts (ferric chloride, alum, etc.), bioadditives, and enzymes. What chemicals are used is determined by the desired outcome. For example, typical pretreatment problems that are treated chemically include reducing odors, neutralizing acids or bases, reducing corrosion, reducing BOD_5, improving solids and grease removal, reducing loading in the plant, and helping along subsequent processes. Attention to ensure mixing is essential to successful chemical pretreatment.

14.2.5.1 Chemical Feeders

Chemical feeders are available in two types: *dry feeders*, which apply dry or powdered chemicals and liquid or *solution feeders*, which apply chemical in solution or suspension.

The two most common types of dry feeders are *volumetric* and *gravimetric*. Choosing which type to select for a particular application depends on whether the chemical is measured by volume (volumetric-type) or weight (gravimetric-type). Volumetric dry feeders are simpler and less expensive than gravimetric pumps, but they are also less accurate. Gravimetric dry feeders are extremely accurate and can deliver high feed rates.

Liquid feeders are generally small positive-displacement metering pumps. *Positive-displacement pumps*, normally used in high pressure, low-flow applications, deliver a specific volume of liquid for each stroke of a piston or diaphragm. Three types are common: (1) reciprocating (piston-plunger or diaphragm feeders), (2) vacuum type (gas chlorinators), or (3) gravity feed rotameter (drip feeders).

14.2.6 FLOW EQUALIZATION

Wastewater treatment plants are subject to wide swings in influent levels and organic loading. Flow equalization directs excessive flow into storage basins, maintaining adequate mixing and aeration during storage for the wastestream to control odor and solids buildup during storage. The stored influent is then pumped back into the treatment process during low flow periods. Flow equalization reduces or removes wide flow rate swings, preventing flows above maximum plant hydraulic capacity, and reducing diurnal flow variations. Equalized flows allow treatment plants to perform optimally by providing stable hydraulic and organic loading.

SUMMARY

The pretreated wastestream is stripped of much of its organic solids in the next treatment step: primary sedimentation.

CHAPTER 15

Primary Sedimentation

15.1 PROCESS PURPOSE AND METHOD

THOUGH screening, comminuting, and grit removal take away much of the wastestreams solids mass, after preliminary treatment the influent still contains suspended organic solids. These settleable organic and floatable solids are readily concentrated and removed by primary/plain sedimentation (or clarification) under relatively quiescent conditions (see Figure 15.1). Primary clarification can be expected to remove 90 to 95% of settleable solids, 40 to 60% of total suspended solids, and 25 to 35% of BOD_5.

Sedimentation is used in primary, secondary, and advanced wastewater treatment processes to remove solids. Primary sedimentation occurs in either long rectangular tanks or circular tanks, usually called primary clarifiers. After wastewater enters a settling tank or basin, velocity reduces to about one foot per minute. Within these basins, the heavier primary settled solids, settle to the bottom. The primary settled solids are removed as sludge, and are generally pumped to a sludge-processing area.

Solids that are lighter than water (oil, grease, and other floating materials) float to the top, forming scum. These floating solids are skimmed from the surface and removed. Wastewater flow leaves the sedimentation tank over an effluent weir for further treatment. Process efficiency is controlled by detention time (normally about two hours), temperature, tank design, and equipment condition.

15.2 PROCESS EQUIPMENT: SEDIMENTATION TANKS

The sedimentation tanks or clarifiers commonly used in primary sedimentation include septic tanks, two story (Imhoff) tanks, and plain settling tanks.

169

Flow from primary settling tank.

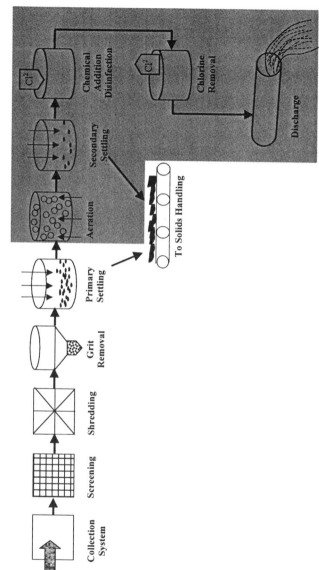

Figure 15.1 Unit processes for wastewater treatment: pretreatment.

171

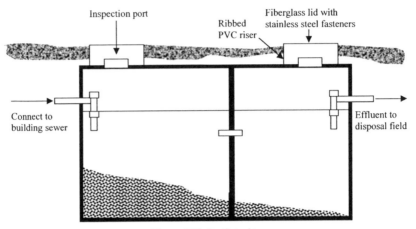

Figure 15.2 Septic tank.

Secondary and advanced wastewater sedimentation processes normally use plain settling tanks.

15.2.1 SEPTIC TANKS

Septic tanks combine settling, skimming, and unmixed anaerobic digestion in one prefabricated unit. Small facilities may use septic tanks for sedimentation, but long detention time and no control for solids separation means that they should not be used for larger applications. Since the decomposing solids are not separated from the wastewater flow, when the solids fill the tank, they overflow into the discharge stream (see Figure 15.2).

15.2.2 TWO STORY TANKS

In two story (Imhoff) tanks, the problem of solids separation in septic tanks is solved. Imhoff tanks consist of a settling compartment for sedimentation, a lower compartment for solids collection and digestion, and gas vents. Slots in the bottom of the settling compartment allow solids to pass into the lower collection chamber, where they decompose anaerobically. Digestion gases are released through the settling compartment vents.

15.2.3 PLAIN SETTLING TANKS

The optimum settling process tanks, plain settling tanks (or clarifiers) can accomplish sludge removal either continuously or intermittently (see opening photo).

As influent enters the tank, it is slowed and distributed evenly across the width and depth of the unit. After a detention time of from 1 to 3 hours (2-hour average), wastewater passes through the unit, and leaves over the effluent weir. The settled sludge is removed for further processing. In continuous operation (which may produce sludge with solids percentages of less than 2 to 3%), the sludge may require further dewatering processes to remove excess water. In intermittent sludge removal, the sludge should settle long enough to obtain 4 to 8% solids, but should be removed frequently enough that clumps of solids don't rise to the surface.

Mechanical scum removal on the tank surface should occur frequently enough to prevent excessive buildup and scum carryover.

Housekeeping requirements for settling tanks include keeping baffles (used to prevent scum carryover), scum troughs, scum collectors, effluent troughs, and effluent weirs free from heavy biological growth and solids accumulation.

Performance of the settling process is evaluated through process control sampling and testing, including tests for settleable solids, dissolved oxygen, pH, temperature, total suspended solids, and BOD_5, as well as sludge solids and volatile matter testing.

15.3 SETTLING TANK EFFLUENT

After preliminary treatment and primary sedimentation, the wastestream (cleaned of large debris, grit, and many settleable materials) is now called *primary effluent*. Primary effluent still contains large amounts of dissolved food, waste, and other chemicals (nutrients), and is usually cloudy and gray in color.

SUMMARY

Most of the nutrients left in primary effluent are removed in secondary treatment processes.

CHAPTER 16

Biological Treatment

16.1 PROCESS PURPOSE

$\mathbf{G}$ RAVITY is the moving force in primary treatment. Primary treatment unit processes (screening, degritting, and primary sedimentation) remove pollutants that either float or settle out of the effluent, leaving behind about 50% of the raw pollutant load. Biological activity is the moving force behind secondary treatment. Secondary treatment includes methods that use biological processes to convert dissolved, suspended, and colloidal organic wastes to more stable solids, which can either be removed by settling or harmlessly discharged to the environment. *Biological treatment* (sometimes called secondary treatment, but in practice, biological treatment is part of secondary treatment processes) provides BOD removal well beyond the levels that primary treatment can achieve, producing an effluent with not more than 30 mg/L BOD_5 and 30 mg/L total suspended solids to meet Clean Water Act (CAA) requirements (see Figure 16.1).

Microorganisms can convert organic wastes (via biological treatment) into stabilized, low-energy compounds. Three commonly used approaches take advantage of this. *Trickling filters*, *rotating biological contactors* (*RBC*), or the *activated sludge* process follow normal primary treatment sequentially. The third treatment process, *ponds* (oxidation ponds or lagoons), however, can provide equivalent results without preliminary treatment.

Most biological treatment processes decompose solids aerobically, producing carbon dioxide, stable solids, and more organisms. All of the biological processes must include some form of solids removal (usually settling tanks or filtration), since the processes produce solids.

Ponds, trickling filters, and rotating biological contactors have been successfully used for wastewater treatment for several years. Trickling filters, for example, have been used since the late 1800s. These systems have their own

175

Distributor arm on trickling filter.

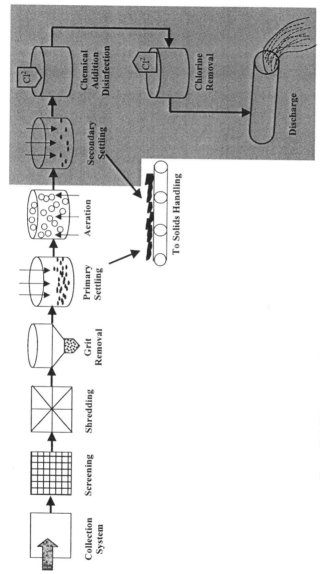

Figure 16.1 Unit processes for wastewater treatment: secondary or biological treatment.

177

unique problems, as well as problems in common. Ponds, trickling filters and RBCs are temperature sensitive, can remove less BOD, and in some cases, can cost more to build than the more modern activated sludge systems, though activated sludge systems may cost more to operate because of energy requirements for running pumps and blowers.

16.2 PROCESS SYSTEMS

Biological treatment processes fall into two large categories: fixed film systems and suspended growth systems.

Fixed film systems [trickling filters and rotating biological contactors (RBCs)] use biological growth (biomass or zoogleal slime) that forms on a media. The media provides a large area for slime growth, as well as ventilation (see Photo 16.1). Wastewater passes over and around the slime on the media. With contact between the wastewater and slime, microbial organisms remove and oxidize the organic solids.

Suspended growth systems use biological growth that mixes with the wastewater. Typical suspended growth systems are various modifications of the activated sludge process.

16.3 TRICKLING FILTERS

Trickling filter systems usually follow primary treatment. They generally include a secondary settling tank or clarifier. Widely used for the treatment of domestic and industrial wastes, the trickling filter biological treatment process is a fixed film method designed to remove BOD_5 and suspended solids.

Wastewater containing organic contaminants contacts a population of microorganisms attached (or fixed) to the surface of filter media. Fist-sized stone or redwood, synthetic materials such as plastic, or any other substance capable of withstanding weather conditions for many years are typically used as filter media (see Figure 16.2).

The primary effluent is disbursed over the top of the filter media by a fixed distributor system or a rotating distribution arm. The wastewater forms a thin layer as it flows down through the filter media and over the microorganism layer on the media surfaces. As the distributor arm rotates (see Photo 16.2), a flow of wastewater and air alternates contact with the microorganism layer (the zoogleal slime). The spaces between the media allow air to circulate easily, maintaining aerobic conditions. The biological slime absorbs and consumes the wastes trickling through the media bed, while the organisms aerobically decompose the solids, producing more organisms and stable wastes. These wastes either become part of the slime or return to the wastewater flowing over the media.

Zoogleal slime is mostly comprised of bacteria, but may also contain algae, protozoa, worms, snails, fungi, and insect larvae. As the slime accumulates,

Photo 16.1 Filter media.

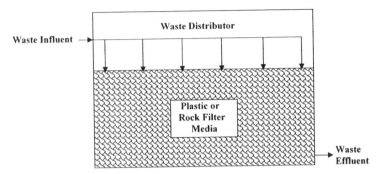

Figure 16.2 Trickling filter.

it builds up and occasionally sloughs off the media materials. The *sloughings* collect at the bottom of the trickling filter with the treated wastewater, and pass on to the secondary settling or sedimentation tank for removal. In a fixed distributor system, the wastewater flow cycles on and off at a specific dosing rate, ensuring that the microorganisms receive an adequate oxygen supply. A trickling filter system's overall performance depends on hydraulic and organic loading, temperature, and recirculation.

Trickling filter systems are typically *roughing filters*, used to reduce organic loading on a downstream activated sludge process. In domestic wastewater treatment, trickling filters are often used for treating wastestreams that produce sludge with poor settling characteristics and poor compactibility (bulking sludge), because the microbial film that sloughs off the trickling filter is relatively dense and can be readily removed by sedimentation.

Trickling filters are also used for the treatment of industrial wastes, including organic chemicals and plastics, and by synthetic fiber industries to treat aqueous waste containing such contaminants as toluene and ethylbenzene.

16.3.1 TRICKLING FILTER PERFORMANCE CLASSIFICATIONS

Trickling filter classifications (hydraulic and organic loading capabilities) determine the expected performance and the trickling filter construction. Classification types include standard rate, intermediate rate, high rate, super high rate (plastic media), and roughing rate types. Currently, standard rate, high rate, and roughing rate are the filter types most commonly used.

- *standard rate filter:* hydraulic loading (gpd/ft^3) of 25 to 90, seasonal sloughing frequency, no recirculation, typical 80 to 85% BOD$_5$ removal rate, and 80 to 85% TSS removal rate
- *high rate filter:* hydraulic loading (gpd/ft^3) of 230 to 900, continuous sloughing frequency, employs recirculation, typical 65 to 80% BOD$_5$ removal rate, and 65 to 80% TSS removal rate

Photo 16.2 Trickling filter media.

181

- *roughing filter:* hydraulic loading (gpd/ft³) of >900, continuous sloughing frequency, recirculation not normally included, typical 40 to 65% removal rate, and 40 to 65% TSS removal rate

16.3.2 TRICKLING FILTER EQUIPMENT

Distribution systems evenly spread wastewater over the entire media surface. *Rotary distributors*, that move above the media surface and spray the surface with wastewater, are the most common systems. Rotary systems are driven by the force of the water leaving the distributor arm orifices. The distributor arms usually have small plates below each orifice that spray the wastewater in a fan-shaped distribution pattern. In *fixed nozzle* systems (frequently used with deep bed synthetic media filters), the nozzles remain in place above the media, and are designed to spray effluent over a fixed portion of the media.

The primary consideration for media selection is that it be capable of providing the desired film locations for biomass development. Media may be 3 to 20 or more feet in depth, depending on the type of media used and the filter classification.

Trickling filters that use ordinary rock are usually only about three meters deep (see Photo 16.3). The weight of the rocks causes structural problems, and also requires wide bed construction—in many cases up to 60 feet in diameter. Lighter weight synthetic media allows much greater bed depths.

The *underdrains* support the media, collect the wastewater and sloughings and carry them out of the filter, and provide ventilation to the filter. The underdrains should never be allowed to flow more than 50% full of wastewater. This ensures sufficient airflow to the filter.

Effluent channels carry the flow from the trickling filter to the secondary settling tank.

The *secondary settling tank* provides two to four hours of detention time to separate the sloughing materials from the treated wastewater. Though design, construction, and operation are similar to primary settling tank design and usage, sloughings are lighter and settle more slowly than does sludge, requiring longer detention times.

Recirculation pumps and *piping* recirculate a portion of the effluent back into the filter influent. This improves the trickling filter or settling tank performance. Systems that include recirculation must also involve pumps and metering devices.

The trickling filter effluent collects in the underdrain system, then travels to a sedimentation tank called a *secondary clarifier*. Secondary clarifier (or final clarifier as it is sometimes called) construction is similar in most respects to the primary clarifier, although differences occur in operation that can include detention time, surface settling rate, hydraulic loading, sludge pumping, overflow rate, weir loading, and other details.

Photo 16.3 Trickling filter.

183

16.4   ROTATING BIOLOGICAL CONTACTOR (RBC)

The rotating biological contactor (RBC) is a biological treatment system that provides an alternative method for the attached growth idea used by trickling filters. RBCs also rely on microorganisms growing on a media surface. The RBC is a *fixed film* biological treatment device, and the basic biological process is similar to that occurring in the trickling filter.

RBCs are comprised of a series of closely-spaced circular plastic (synthetic) disks mounted side by side, and usually about 3.5 m in diameter. These disks are attached to a rotating horizontal shaft. In use, each disk unit is set up so that about 40% of each disk surface is submerged in a tank containing the wastewater for treatment. The RBC unit rotates slowly, and the biomass film (zoogleal slime) that grows on the disk surfaces is carried in and out of the wastewater. The microorganisms that form the portion of the biomass that is submerged in wastewater absorb organics; when they rotate out of the wastewater, they receive the oxygen needed for aerobic decomposition. When the zoogleal slime reenters the wastewater, any excess solids and waste products slip off the media as sloughings, which are transported with the effluent to a settling tank for removal.

Modular RBC units are placed in series (individual contactor units are *stages*; the group is a *train*) because a single contactor cannot achieve the required treatment levels; with a properly designed and operated RBC, the treatment achieved can exceed conventional secondary treatment.

RBC systems are usually made up of two or more trains consisting of three or more stages in each. RBCs are easier to operate under varying load conditions than are trickling filters, since keeping the solid medium wet at all times is easier. With multiple-stage RBC systems, the achievable level of nitrification is significant (see Figure 16.3).

RBCs are used to treat dilute aqueous wastes that contain biodegradable organics, including solvents. RBC systems can withstand organic and hydraulic surges effectively, because they possess a large biological cell mass to handle the loading. RBCs allow a greater degree of control over treatment variables than trickling filters. The microorganisms, the organic waste, and atmospheric oxygen—and their rate of contact—can be controlled by adjusting the submerge depth for the disks, and the disk rotation rate. RBC operational problems can include central shaft deflection and difficulty in controlling growth (the slime layer tends to overgrow or slough off completely).

16.4.1   RBC EQUIPMENT

An RBC is made up of the rotating biological contactor with disks built out of either standard or high density media mounted on a center shaft, a drive system, a tank, baffles, a housing or cover, and a settling tank (see Figure 16.4).

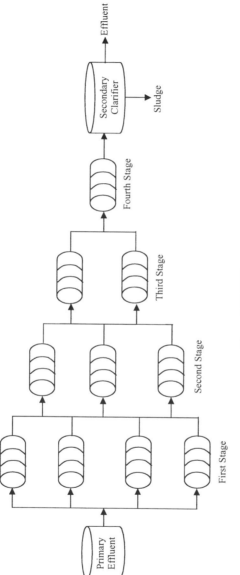

Figure 16.3 RBC system.

185

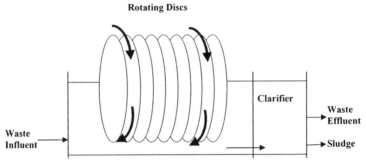

Figure 16.4 Rotating biological contactor.

The rotating biological contactor disks (the media) contain large surface areas for biomass growth.

The *center shaft* supports the media. While shafts are designed to support the weight of the media and the biomass, a major problem has been support shaft collapse.

The *drive system* rotates the shaft and disks. Drive systems can be air driven, mechanical, or a combination of both. Major operational problems can arise if the drive system does not provide uniform RBC movement.

The *tank* holds the wastewater that the RBC rotates in, and should possess enough volume to permit depth and detention time variations.

Baffles permit proper adjustment of the loading applied to each stage of the RBC process. Baffles should allow for adjustment to increase or decrease RBC submergence.

Normally RBC stages are enclosed in a protective structure (*cover*) to prevent biomass loss from severe weather changes (snow, rain, temperature, wind, and sunlight). Housing can significantly restrict access to the RBC for operational assessment, adjustment, or repair.

The *settling tank* removes the sloughing material created by biological activity. Similar in design to the primary settling tank, the RBC settling tank provides two to four hour detention time to permit settling of lighter biological solids.

Like trickling filters, RBC systems need a secondary clarifier to settle out excess biological solids that slough off the discs as the slime layer thickens. Because microbial metabolic rates slow down with temperature decreases, RBC process efficiency is adversely affected by low temperatures.

16.5 TREATMENT PONDS AND LAGOONS

In terms of construction and operation costs, ponds or lagoons for secondary treatment can provide many advantages for areas where property costs are low and land is available.

Ponds are generally much more simple to build and manage than mechanical systems; large fluctuations in flow generally do not present a problem, and ponds can produce a highly purified effluent at a level of treatment that approaches conventional systems, at much lower cost. In fact, economic concern drives many managerial decisions for the pond/lagoon option.

The level of treatment a pond or lagoon system provides depends on the type and number of ponds used. Ponds can be used as the only secondary treatment, or with other forms of wastewater treatment—that is, other treatment processes followed by a pond or a pond followed by other treatment processes.

Ponds are beginning to be used with increasing frequency in areas where land is readily available, despite potential difficulty in TSS removal efficiency. Construction cost advantages, operation and maintenance advantages, and negligible energy costs make this an attractive option, particularly for small communities.

A stabilization pond treatment system is the simplest to operate and maintain of any of the biological treatment systems. Operation and maintenance activities include collecting and testing samples for dissolved oxygen (DO) and pH, removing weeds and other debris (scum) from the pond, mowing the berms, repairing erosion, and removing burrowing animals.

16.5.1 POND TYPES

Ponds can be classified (named) based on their location in the system, by the wastes they receive, and by the main biological process occurring in the ponds: raw sewage stabilization (RSS) ponds, oxidation ponds, and polishing ponds. Ponds are also classified by the processes that occur within them as aerobic ponds, anaerobic ponds, facultative ponds, and aerated ponds.

16.5.2 PONDS BY LOCATION AND WASTE

16.5.2.1 Raw Sewage Stabilization Pond

Raw sewage stabilization (RSS) ponds are the most common type of pond. Effluent placed in these ponds receives no prior treatment except screening and shredding. Raw sewage stabilization ponds usually provide a minimum of 45 days detention time, and receive no more than 30 pounds of BOD_5 per day per acre. Discharge quality depends on the time of the year. In summer, RSS ponds produce high BOD_5 removal and excellent suspended solids removal.

The pond consists of an influent structure, pond berm or walls, and an effluent structure designed to permit effluent selection for quality. An RSS pond's normal operating depth is three to five feet.

Bacteria decompose the organics in the wastewater (aerobically and anaerobically) and algae use the products of the bacterial action to produce oxygen (photosynthesis) in the biological processes that occur in RSS ponds.

When wastewater enters a stabilization pond, settling, aerobic decomposition, anaerobic decomposition and photosynthesis all begin to occur. The solids that enter the pond in the wastewater, as well as the solids produced by the biological activity, all settle to the bottom. Over the course of the pond's use, this eventually reduces detention time and pond performance, and the pond must be replaced or cleaned. This time span is usually 20 to 30 years.

Bacteria and other microorganisms ingest the organic matter as their primary food source. Through aerobic decomposition, they use oxygen, organic matter, and nutrients to produce carbon dioxide, water, and stable solids that can settle out, as well as more organisms. The carbon dioxide they produce feeds the photosynthesis process that occurs near the pond surface.

In the lower levels of the pond, organisms also use the settled-out solids as food material. The oxygen levels at the bottom of the pond are extremely low so the process used is anaerobic decomposition, producing gases (hydrogen sulfide, methane, etc.) that dissolve in the water, stable solids, and more organisms.

A population of green algae develops near the pond surface. Using the carbon dioxide produced by the bacterial population, nutrients, and sunlight, the algae growth produces more algae and oxygen that is dissolved into the water. The dissolved oxygen is then used by organisms in the aerobic decomposition process.

16.5.2.2 Oxidation Ponds

Oxidation ponds (usually designed from the same criteria as stabilization ponds) receive flows from stabilization ponds or primary settling tanks. These ponds provide biological treatment, additional settling, and some reduction of fecal coliform levels.

16.5.2.3 Polishing Ponds

Polishing ponds use the same equipment as a stabilization ponds. They receive flow from oxidation ponds or other secondary treatment systems. Polishing ponds remove additional BOD_5, solids and fecal coliform, and some nutrients. They generally provide 1 to 3 days detention time and operate at depths from 5 to 10 feet. If the wastewater for treatment is held in the pond too long, or the pond is too shallow, algae growth will begin, which causes increased influent suspended solids concentrations.

16.5.3 PONDS BY POND BIOLOGICAL PROCESS

16.5.3.1 Aerobic Ponds

In aerobic ponds, oxygen is present throughout the pond. All biological activity is aerobic decomposition. These ponds are not widely used.

16.5.3.2   Anaerobic Ponds

In anaerobic ponds, no oxygen is present in the pond, and all biological activity is anaerobic decomposition. These ponds are normally used to treat high-strength industrial wastes.

16.5.3.3   Facultative Ponds

Facultative ponds are the most common pond type by process. In the upper portions of the pond, the presence of oxygen means aerobic processes occur. In the lower levels of the pond, no oxygen is present and anoxic and anaerobic processes occur.

16.5.3.4   Aerated Ponds

In aerated ponds, mechanical or diffused air systems provide oxygen throughout the pond. With aeration, pond depth and/or the acceptable loading levels may increase. Mechanical or diffused aeration can supplement or replace natural oxygen production.

16.6   ACTIVATED SLUDGE SYSTEMS

Currently the most widely used biological treatment, the activated sludge process recirculates part of the biomass as an integral part of the process. This allows relatively short acclimation processes for microorganism adaptation to changes in wastewater composition, and a greater degree of control over the acclimated bacterial population.

16.6.1   ACTIVATED SLUDGE PROCESS OPERATION

The activated sludge process removes BOD_5 and suspended matter through aerobic decomposition. With proper process controls adjustment, nitrogen and phosphorous may also be removed.

All activated sludge systems include an aeration basin followed by a settling tank (see Figure 16.5). The aeration tank receives effluent from the primary clarifier, as well as a mass of recycled biological organisms from the secondary settling tank—the *activated sludge*. Air or oxygen is pumped into the tank via blowers to maintain aerobic conditions, and the effluent, oxygen and activated sludge are kept thoroughly agitated by mixers for about six to eight hours. The wastewater (*mixed liquor*) then flows into the secondary settling tank, where the solids (mostly bacterial masses) settle from the liquid by subsidence. Some of the solids are returned to the aeration tank to maintain

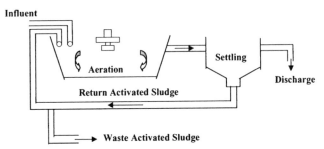

Figure 16.5 Activated sludge process (simplified).

the proper bacterial population there, while the remainder is processed and disposed.

Factors that affect the performance of an activated sludge system include temperature, return rates, amount of oxygen available, amount of organic matter available, pH, waste rates, aeration time, and wastewater toxicity.

Obtaining desired performance levels in an activated sludge system means a proper balance must be maintained between the amount of food (organic matter), organisms (activated sludge), and oxygen [dissolved oxygen (DO)]. Most problems with activated sludge systems are caused by an imbalance among these three items.

16.6.2 ACTIVATED SLUDGE PROCESS EQUIPMENT

More complex than for the other processes discussed, equipment for activated sludge treatment processes includes an aeration tank, aeration system, system settling tank, and return sludge and waste sludge systems (see Figure 16.6).

Aeration tanks provide the required detention time, which depends on the specific modifications. Detention time ensures that the activated sludge and the influent wastewater are thoroughly mixed. Tank design normally attempts to ensure that all tank areas are thoroughly aerated, leaving no dead spots.

Both mechanical and diffused aeration are common. *Mechanical aeration* systems use agitators or mixers to combine air and mixed liquor, or sparge rings to release air directly into the mixer

Diffused aeration systems release pressurized air through diffusers near the tank bottom. The size of the air bubbles produced directly affects system efficiency—the finer the bubbles, the higher the efficiency levels that can be achieved. Diffused air systems have blowers that produce large volumes of low pressure air (5 to 10 psi), lines that carry the air to the aeration tank, and headers that distribute the air to the diffusers, which release the air into the wastewater (see Photos 16.4 and 16.5).

Activated sludge systems require plain *settling tanks* to provide two to four hours of hydraulic detention time.

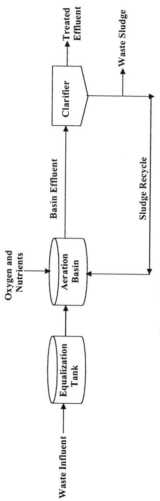

Figure 16.6 Activate sludge process.

Photo 16.4 Aeration basin.

Photo 16.5 Aeration basin.

Return sludge systems include pumps, a timer or variable speed drive to regulate pump delivery, and a flow measurement device to determine actual flow rates.

Waste activated sludge withdrawal is sometimes accomplished by adjusting valves on the return system. For a separate system, equipment requirements include pumps, a timer or variable speed drive, and a flow measurement device.

SUMMARY

When the biological treatment unit for treating wastewater is a trickling filter, rotating biological contactor, or activated-sludge system, these unit treatment processes must be followed by secondary sedimentation to remove the accumulated biomass. Secondary sedimentation is covered in Chapter 17.

Secondary Sedimentation

17.1 PROCESS PURPOSE

S ECONDARY sedimentation immediately follows biological treatment, and is required before any advanced treatment processes can occur (see Figure 17.1). Sedimentation must also occur before disinfection prior to effluent discharge to the receiving water body.

The effluent from these systems contains considerable biomass—a significant organic load—that must be removed by secondary sedimentation to meet acceptable effluent standards. In secondary sedimentation, the effluent is sent to a sedimentation tank called a *secondary clarifier* or *final clarifier*. Similar to the primary sedimentation tank or primary clarifier, differences occur in detention time, overflow rate, and weir loading.

Activated sludge processes create high solids loading and fluffy biological floc. Operationally, secondary sedimentation units perform two important functions.

- They separate the mixed-liquor suspended solids from the treated wastewater, resulting in an effluent sufficiently clarified to meet regulatory standards.
- They concentrate or thicken the return sludge to minimize the quantity of sludge that must be handled.

17.2 SECONDARY SETTLING

The function secondary sedimentation performs is essential to the overall treatment process. In the operation of a typical activated-sludge unit process, for example, organic pollutants are absorbed by the millions of microorganisms

195

Secondary sedimentation (Lancaster, PA Advanced Wastewater Treatment Plant).

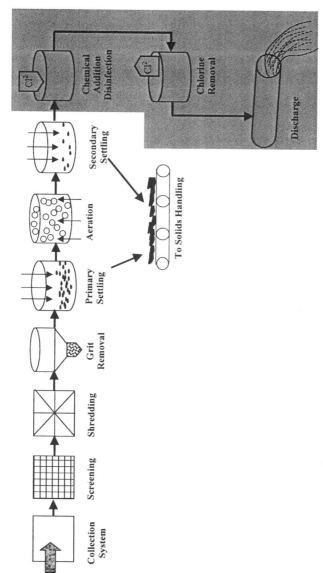

Figure 17.1 Unit processes for wastewater treatment: secondary sedimentation.

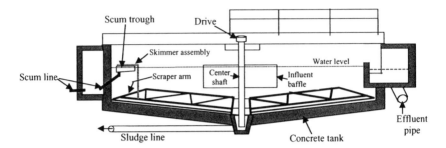

Figure 17.2 Sedimentation tank.

(the activated sludge) in an aeration tank, converting the contaminants to more stable forms of biomass (sludge). This activated-sludge conversion process is only effective if proper clarification or separation of the sludge from the liquid portion of the mixed liquor occurs. The secondary sedimentation tank (or secondary clarifier) provides the environment whereby separation via gravity settling can take place (see Figure 17.2).

The critical process of gravity settling must be properly controlled and maintained. Unsettled sludge carries over the clarifier's effluent weirs and could contaminate the receiving body of water. The solids concentration within the secondary sedimentation basin is the key parameter to consider. Solids concentration is generally considered equal to the solids concentration in the aeration tank effluent. Solids concentrations can also be determined in the laboratory using a secondary clarifier core sample.

Gravity settling is the most important part of the activated sludge treatment system. Ensuring optimum operation—the removal of 90+% of the raw sewage BOD and TSS—means certain sedimentation tank observations are needed to maintain proper system control. Sedimentation tank observations include flow pattern to ensure normal uniform distribution; settling, normal to very low amounts and type of solids leaving with the process effluent; and the usual very clear (low turbidity) process effluent.

SUMMARY

Sedimentation is essential at many stages of water and wastewater treatment, where settling prepares the wastestream for further treatment processes. After secondary sedimentation, some wastewaters are ready for disinfection and discharge, and other wastewaters require advanced treatment.

CHAPTER 18

Advanced Treatment

18.1 PROCESS PURPOSE

S ECONDARY wastewater treatment coupled with disinfection normally re- moves 85+% of the BOD and suspended solids, and nearly all the pathogens. However, even a properly operating secondary treatment facility achieves only minor removal of nitrogen, phosphorus, soluble COD, and heavy metals. Ad- vanced wastewater treatment (or AWT) adds unit processes that remove more contaminants from wastewater than can usually be achieved by primary and secondary treatment.

Advanced or tertiary treatment is employed to meet specific parameters of discharge or reuse criteria. Nitrogen, phosphorus, soluble COD, and heavy met- als are of major concern under some circumstances, especially when discharge requirements for a particular area may be more stringent than effluent from secondary treatment can achieve.

Each body of water has its own particular characteristics. Each reacts some- what differently to the discharge of treated wastewater. Conventional secondary treatment is usually adequate to make the discharge safe for the receiving stream, but sometimes higher degrees of treatment must be provided. In these instances, advanced or tertiary treatment is needed. Under circumstances that include ef- fluent discharge into delicate ecosystems, or discharge of large amounts of effluent into small receiving bodies, installing systems capable of removing these pollutants to a greater degree is the option usually chosen. These ad- vanced or tertiary treatment processes improve the effluent quality to a level useful for many reuse purposes—but considerations and costs are involved.

Decisions to install advanced treatment processes are not made lightly. Usu- ally expensive to build and operate, advanced systems also normally demand a highly trained operating crew. Sometimes the sludges they produce are difficult

199

Phosphorus testing (Lancaster, PA Advanced Wastewater Treatment Plant).

to dispose of economically. These techniques are installed (with need usually determined on a case-by-case basis) to tackle the toughest kinds of wastewater problems. The end result of advanced treatment produces treated water pure enough for industrial process uses. Different levels of advanced treatment produce effluent that can be used for golf course and other landscape irrigation, or to replenish groundwater.

While not all wastewater effluents are presently being treated beyond secondary treatment for beneficial reuse, increasing population levels are applying pressure for more and considered wastewater recycling and reuse. Wastewater reuse at every level will soon have to be common practice, especially in areas with low rainfall or uncertain water supply.

That we are already indirectly reusing wastewater is not always obvious to everyone. Discharging wastewater to watercourses has been practiced for many years. Communities downstream on watercourses use them as sources of water supply. Urban population growth causes problems associated with more people creating more sewerage discharge of both treated and untreated wastewaters. Downstream populations are *reusing* that water—heavily diluted, we hope. This practice is a form of indirect wastewater reuse.

18.2 ADVANCED TREATMENT PROCESSES

Some of the more common tertiary treatment processes and operations include effluent polishing, nitrogen and phosphorus removal, and land application.

18.2.1 EFFLUENT POLISHING

Tertiary treatment is often designed to "polish" the final effluent by removing BOD and TSS—additional suspended solids—mostly organic compounds. Polishing is generally accomplished by implementing a filter (usually granular-media type), much like filters used for drinking water purification.

Gravity filtration in an open tank, and filtration under pressure in closed pressure vessels are common filtration methods used for effluent polishing. Whatever method is used, special care must be taken with the filtration unit itself. The presence of these organic and biodegradable suspended solids in the secondary effluent means that tertiary filters must be backwashed frequently, usually followed by an auxiliary surface air-wash to thoroughly scour and clean the filter bed. Without frequent backwashing, decomposition causes the filter bed to develop septic or anaerobic conditions.

In *microstraining*, another method of effluent polishing, microstrainers or microscreens composed of specially woven steel wire cloth are mounted around the perimeter of a large revolving drum. Partially submerged in secondary effluent, that flows into the drum and then outward through the microscreen, the drum rotates, capturing solids that the rotation carries to the top. There a water spray flushes them into a hopper.

18.2.2 NITROGEN REMOVAL

When nitrogen is present in wastewater, it can appear in soluble form as organic nitrogen, ammonia, or nitrate compounds. Nitrogen compound removal is needed for several reasons, including nutrient removal to help control algal blooms in the receiving body. The ammonia forms of nitrogen can be toxic to fish, and removal prevents additional oxygen demand in receiving waters, as nitrogen is converted to nitrate.

Wastewater nitrogen is removed using either a biological process (*nitrification/denitrification*) or by a chemical process (*ammonia stripping*).

Nitrification/denitrification is a two step process. In the nitrifying step, secondary effluent enters an additional aeration tank or other biological unit process (a trickling filter, for example), where nitrifying bacteria thrive. These microorganisms convert ammonia nitrogen to nitrate nitrogen, a form of nitrogen that is not toxic to fish, and that does not cause an additional oxygen demand. In the second step (denitrification), different bacteria anaerobically convert nitrates to nitrogen gas (N_2).

In the ammonia stripping process, treated wastewater's pH is raised to at least 10, usually using quick lime (CaO). This forms dissolved ammonia gas, which is then freed from the effluent in a stripping tower. Ammonia stripping is generally more cost-effective than biological nitrification/denitrification, but has limitations. The lime added reacts with carbon dioxide in air and water, forming calcium carbonate scale, which must be periodically removed. In cold-weather conditions, air stripping loses efficiency. Low temperatures cause problems with icing and reduced stripping ability caused by the increased solubility of ammonia in cold water (Davis and Cornwell, 1991). The air stripping process also simply transfers the pollution problem from water to air, creating an additional burden on the atmosphere (Masters, 1991).

18.2.3 PHOSPHORUS REMOVAL

Biological treatment unit processes used in wastewater treatment only remove about 30% of the phosphorus in municipal wastewater. This 30% removal has become unacceptable in many areas because phosphorus becomes a limiting nutrient when released to the receiving body, sometimes leading to increased eutrophication.

Phosphorus removal usually involves chemicals (e.g., alum, ferric chloride, or lime) added to the wastewater at some point in the conventional process, avoiding the need for additional tanks and filters (for more information on chemical feeders, see Section 14.5.2.1). Through *chemical precipitation* of the

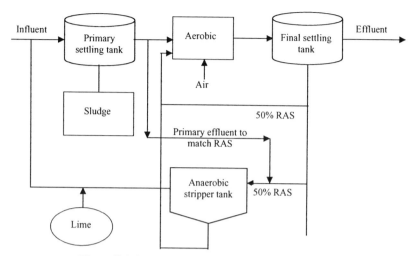

Figure 18.1 Sidestream process for biophosphorus removal.

phosphate ions and coagulation, the organic phosphorus compounds become trapped in coagulant flocs and settle out in a clarifier (see Figure 18.1).

18.2.4 MEMBRANE PROCESSES FOR ADVANCED TREATMENT

Semipermeable membranes are used to separate various contaminants and impurities from wastewater in *reverse osmosis*, by permeating high-quality water and halting the passage of dissolved solids. Some of the constituents and contaminants that reverse osmosis can remove include arsenic, asbestos, atrazine, fluoride, lead, mercury, nitrate, radium, and volatile contaminants such as benzene, trichloroethylene, trihalomethanes, and radon.

In osmosis, water naturally passes from the weaker solution to the stronger solution, equalizing the chemical balance in the membrane-separated solutions. Osmotic pressure drives osmosis. In reverse osmosis, external pressure stronger than the natural osmotic pressure causes water to reverse flow from the natural osmotic direction through the membrane.

Pretreatment is essential to effective reverse osmosis. Organics, microorganisms, oil, and grease will coat and foul the membranes, and scale-forming constituents (calcium, iron, manganese, silicon, etc.) will damage and reduce membrane capability.

Equipment for reverse osmosis includes pretreatment equipment, the membrane structure and support, a high pressure pump and piping, and a brine-handling and disposal system.

In *electrodialysis*, water is demineralized using ion-selective membranes and an electric field to separate anions and cations in solution. Electrodialysis is finding increasing use in industrial waste treatment. Metals salts from plating rinses are sometimes removed by electrodialysis, although the most common use for the technique remains the treatment of brackish waters.

When an electric charge is applied to influent in cells (stacks) containing alternatively arranged cation- and anion-permeable membranes, the cations and anions in the influent migrate to the opposite poles. There they pass through the selective membranes and separate into pure and brine streams.

Electrodialysis systems include a pretreatment system to prevent membrane fouling (as with reverse osmosis), the cellular membrane stack, and a system to handle and dispose the separated brine.

18.2.5 LAND APPLICATION

Land application provides inexpensive and effective tertiary treatment of wastewater, as well as adding the moisture and nutrients needed for vegetation growth, and recharge of groundwater aquifers. Land treatment allows direct beneficial water and nutrient recycling. However, effective wastewater application to land requires relatively large land areas. Determining the feasibility and design of land treatment processes involves assessing the critical factors of soil type and climate.

Soil's filtering characteristics are an effective alternative to expensive and complicated tertiary treatment methods. Natural filtering processes occur as the effluent flows over vegetated ground surface and percolates through the soil. Soil filtration can inexpensively produce a high-quality effluent low in TSS, BOD, phosphorous and nitrogen compounds, and refractory organics. While the basic objective of land application is wastewater treatment and disposal, another goal is to obtain economic and environmental benefits. This procedure can be used to conserve potable water by using secondary effluent to irrigate lawns and other landscaped areas, as well as to produce animal feed crops.

Variations of three basic types or modes of land treatment are common: irrigation or slow rate, infiltration-percolation or rapid infiltration, and overland flow [see Figure 18.2(a), (b), and (c)]. Conditions under which these types function best, and basic objectives of these modes of treatment vary (USEPA, 1977).

- *Irrigation, slow rate land application mode:* wastewater (applied or sprayed to the field surface by ridge-and-furrow surface spreading or sprinkler systems) enters the soil. Vegetation is the critical component for this treatment process. Crops growing on the irrigation area use the available nutrients the effluent contains, while soil organisms stabilize the flow's organic content. The water enters surface water or groundwater, or evaporates [see Figure 18.2(a)].

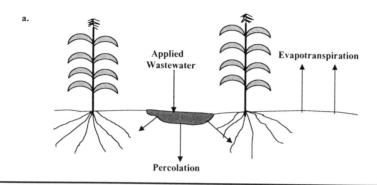

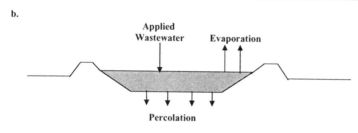

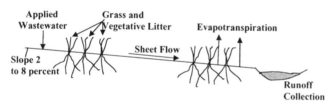

Figure 18.2 (a) Slow land treatment; (b) rapid filtration; (c) overland flow.

- *Infiltration-percolation, rapid infiltration mode:* wastewater pumped to spreading basins begins to evaporate. The remainder infiltrates into the soil. The solids are removed by soil filtration while the remaining water recharges the groundwater system. Soils must be highly permeable for this method to work properly [see Figure 18.2(b)].

- *Overland flow mode:* wastewater (sprayed over sloped terraces) flows slowly over the surface. Physical, chemical, and biological processes work to purify the effluent as the wastewater flows in a thin film down the relatively impermeable surface. Soil and vegetation remove suspended solids, nutrients, and organics, while a small amount of wastewater evaporates. The remaining wastewater flows to collection channels, and the collected effluent is usually discharged to surface waters (see Figure 18.2(a), (b), and (c)].

SUMMARY

After advanced treatment procedures, the wastestream is ready for the final
stage of treatment—disinfection.

REFERENCES

Davis, M. L. and Cornwell, D. A. *Introduction to Environmental Engineering.* New
York: McGraw-Hill, 1991.

Masters, G. M. *Introduction to Environmental Engineering & Science.* Englewood Cliffs,
NJ: Prentice Hall, 1991.

USEPA. *Process Design Manual for Land Treatment of Municipal Wastewater,* 1977.

CHAPTER 19

Wastewater Disinfection

19.1 PROCESS PURPOSE

THE final unit process before effluent discharge into a receiving body, *disinfection*, has several important objectives (see Figure 19.1). Disinfection's main purpose is to protect public health by reducing the organism population in the wastewater to levels low enough to ensure that pathogenic organisms will not be present in sufficient quantities to cause disease when discharged. Disinfection helps prevent the spread of disease and protects our waters: drinking water supplies, beaches, recreational waters, and shellfish growing areas.

Chlorine is the most common disinfectant in use, for both water and wastewater treatment. Alternatives to chlorine use include chlorine dioxide, ozonation, potassium permanganate, ultraviolet (UV) radiation systems, membrane processes, air stripping, and activated carbon adsorption.

The disinfection processes most commonly used for wastewater disinfection are chlorination and dechlorination, ultraviolet irradiation (UV), ozonation, and disinfection by bromine chloride.

19.2 CHLORINATION

Chlorine disinfection has advantages of cost, dependability, and performance predictability (see Table 19.1).

Chlorine is a very reactive substance and readily reacts with other substances, including many chemicals, organic matter, and ammonia. These chemical reactions reduce chlorine, using it up so that it is no longer available for disinfection. The amount of chlorine taken up by organic matter and ammonia in wastewater disinfection is known as *chlorine demand*. When chlorine reacts with ammonia,

207

Chemical metering pump.

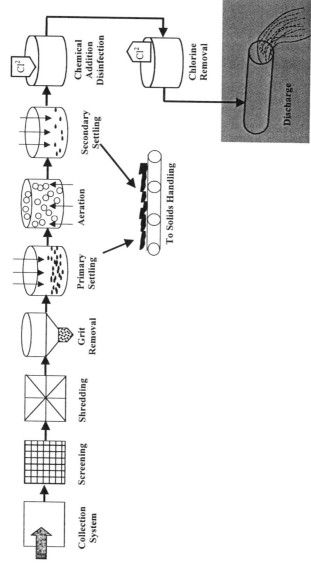

Figure 19.1 Unit processes for wastewater treatment: disinfection and disinfection removal.

TABLE 19.1. Chlorine Disinfection Advantages and Disadvantages.

Advantages	Disadvantages
Relatively inexpensive compared with other disinfection methods	Increased safety and regulatory requirements caused increased operational and liability costs
Dependable chlorination equipment designs reduce breakdowns	Produces chloramines and other substances toxic to fish and aquatic organisms, even in very low concentrations
Process familiarity makes identifying and correcting problems fast and simple	Possible health hazards from chlorine by-products require additional treatment steps
Well documented disinfection performance	Chlorine by-product compounds remain active in the receiving stream for very long periods
Easy-to-use control mechanism (residual and contact time)	Potential in-plant hazards demand safety precautions, personal protective equipment, emergency response plans, training, and regulatory monitoring and reporting

it produces chloramines, that consist of chlorine, nitrogen, and hydrogen. Chloramines possess some disinfecting capability, and are part of the *combined chlorine residual* in chlorine testing.

When all the chemical demands are met, chlorine reacts with water, forming hypochlorous (HOCl) and hydrochloric (Hcl) acids. The most effective chlorine disinfectant, hypochlorous acid is known as *free residual chlorine.*

Chemicals present in the wastewater make achieving free residual impractical for most wastewater treatment plants, so disinfection is usually brought about by combined residual, and controlled by monitoring the *total residual chlorine* (TRC) (see Table 19.2).

Residual level, contact time, and effluent quality all affect disinfection. Chlorine's efficacy decreases with increasing pH and decreasing temperature; it is also affected by ammonia or organic nitrogen. Maintaining the desired residual levels for the required contact time is essential. Failure to do so results in lower

TABLE 19.2. Toxicity Levels of Total Residual Chlorine.

Total Residual Chlorine (TRC) mg/L	
0.06	Toxic to striped bass larvae
0.31	Toxic to white perch larvae
0.5 to 1.0	Typical drinking water residual
1.0 to 3.0	Recommended for swimming pools

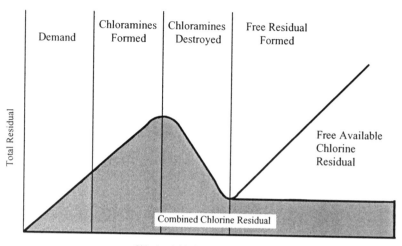

Figure 19.2 Breakpoint chlorination curve.

efficiency and the increased probability of disease organisms in the discharge. For disinfection of most domestic wastes, a total residual chlorine concentration of ≥ 1.0 mg/L should occur for at least 30 minutes at design flow. High levels of solids in the effluent increase residual and/or detention time for proper disinfection (see Figure 19.2).

19.2.1 OTHER FORMS OF CHLORINATION: HYPOCHLORITE

Chlorine substitutes used in disinfection are usually in the form of hypochlorite (similar to the chlorine oxidants used in home swimming pools) or free chlorine gas.

Though hypochlorite presents some minor hazards (skin irritation, nose irritation, and burning eyes), working with it is relatively safe. Available in dry form as powder, pellet, or tablet (calcium hypochlorite), or in liquid form (sodium hypochlorite), hypochlorite can be added directly using a dry chemical feeder, or dissolved and fed as a solution (see Figure 19.3).

19.3 DECHLORINATION

High chlorine concentrations released into the environment can have an adverse environmental impact. Wastewaters heavy in chlorine released from wastewater treatment plant outfall pipes can kill fish and other aquatic life in that area. Many State Water Control Boards have established chlorine water

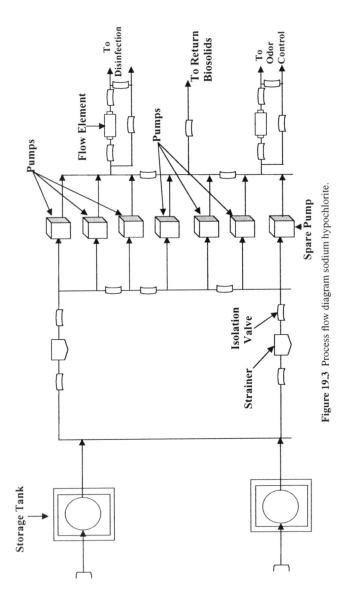

Figure 19.3 Process flow diagram sodium hypochlorite.

quality standards of total residual chlorine μ0.011 mg/L in fresh waters and μ0.0075 mg/L for chlorine-produced oxidants in saline waters.

To comply with these regulations, and to protect aquatic life, sometimes dechlorinating the effluent is necessary. For compliance, many treatment systems have added additional treatment steps for chlorine removal prior to discharge. The dechlorination process uses sulfur dioxide, sodium sulfite, or sodium metabisulfite, any of which react quickly with chlorine and convert it to a less harmful form.

No additional contact tanks are needed, though the dechlorination process does require additional chemical feed and monitoring equipment. Generally, the equipment required for dechlorination is similar to that required for chlorination, and specific equipment depends on the chemical choice. Because chemicals used for dechlorination also react with dissolved oxygen, dechlorinated effluent usually requires aeration prior to discharge.

19.3.1 SULFUR DIOXIDE DECHLORINATION

Dechlorinating with sulfur dioxide involves adding it to the chlorinated wastestream right before outfall. Liquid sulfur dioxide is converted to gaseous form in an evaporator. From the evaporator, the gaseous sulfur dioxide enters the sulfonator, which is similar in design and function to a chlorinator. The sulfonator injects the gaseous sulfur dioxide into the wastestream where it forms sulfurous acid. This is transported to the application site (see Figure 19.4, Photo 19.1 and Photo 19.2). At the application site, sufficient mixing ensures uniform solution distribution in the wastestream. Because reaction time is very short, no contact tank is needed.

Exposure to sulfur dioxide causes damage to the central nervous system. Mixing chlorine gas and sulfur dioxide causes a violent reaction. Switching metering equipment from one gas to the other without first flushing thoroughly with clean, dry air or an inert gas such as nitrogen is extremely dangerous.

19.4 UV (ULTRAVIOLET) RADIATION

Ultraviolet radiation is a highly effective disinfectant under ideal conditions (see Table 19.3).

When UV radiation is used in wastewater treatment, wastewater effluent is exposed to ultraviolet light of a specified wavelength and intensity for a specified contact period. The effectiveness of the process depends on UV light intensity, contact time, and the wastestream turbidity levels.

The effectiveness of UV radiation in wastewater treatment hinges on turbidity. UV lightwaves cannot penetrate solids; this can allow microbial life to survive disinfection. Many states limit UV disinfection usage to facilities that

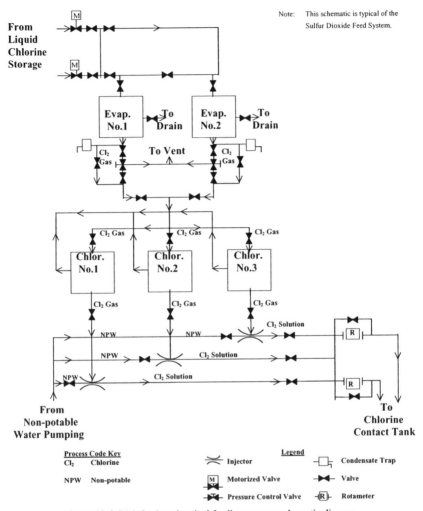

Figure 19.4 Disinfection chemical feeding process schematic diagram.

can regularly produce an effluent containing $\mu 30$ mg/L or less of BOD$_5$ and total suspended solids.

UV disinfection takes place in contact tanks, designed with the banks of UV lights in a horizontal position (either parallel or perpendicular to effluent flow), or with banks of lights placed in a vertical position (perpendicular to flow). The contact tank must provide at a minimum 10-second exposure time (see Figure 19.5).

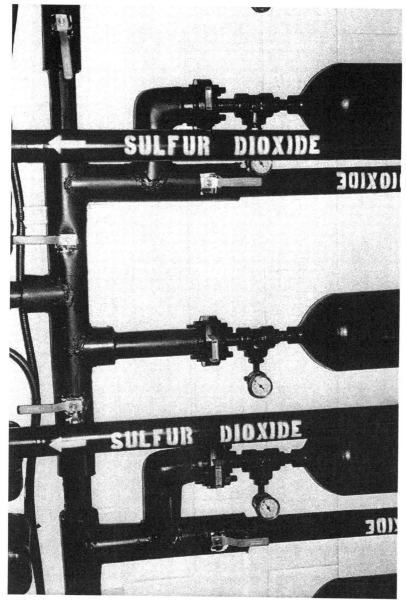

Photo 19.1 Sulfur dioxide piping.

215

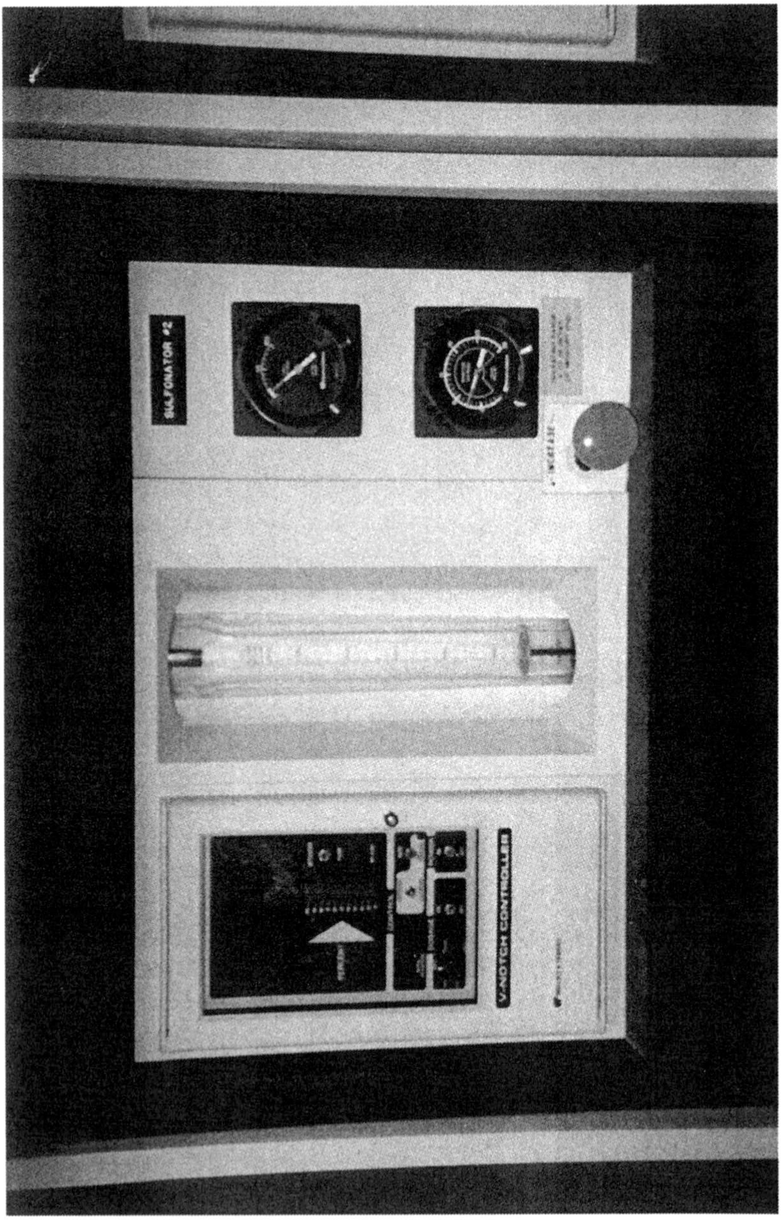

Photo 19.2 Sulfonator control panel.

TABLE 19.3. UV Radiation Advantages and Disadvantages.

Advantages	Disadvantages
Excellent germicidal qualities	Turbidity levels affect UV radiation's ability to disinfect, allowing possible microbial survival
Effectively destroys microorganisms	Maintenance includes regular tube cleaning and replacement as needed. Periodic acid washing removes chemical buildup
Use in hospitals, biological testing facilities, and many other similar locations for sterilization means that effectiveness is well tested	Extremely hazardous to the eyes, requires proper eye protection

19.5 OZONATION

Ozonation provides several advantages over chlorination (see Table 19.4).

Ozone, a strong oxidizing gas, reacts with most organic and many inorganic molecules. Produced when oxygen molecules separate and collide with other oxygen atoms, the ozone molecule is formed of three oxygen atoms. Ozone is an excellent disinfectant for high quality effluents, but less so for turbid wastewaters. For this reason, current regulations for domestic treatment systems limit its use to filtered effluents, unless effluent quality from a system can be proven before installation of the ozonation equipment.

Ozonation unit process equipment includes an oxygen generator as well as an ozone generator. The contact tank requirements include a 10-minute contact time at design average daily flow, and off-gas monitoring for process control. Ozone is extremely toxic and safety equipment capable of monitoring ozone in the atmosphere, as well as a ventilation system that prevents ozone levels higher than 0.1 ppm, are required (see Figure 19.6).

Ozone unit process operation involves ozone generator monitoring and adjustment, and control system monitoring to ensure the required ozone concentrations. Biological testing is used regularly to assess process effectiveness.

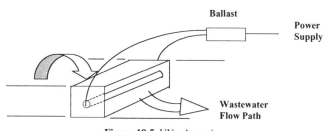

Figure 19.5 UV schematic.

TABLE 19.4. Ozonation Advantages and Disadvantages.

Advantages	Disadvantages
Increases DO in the effluent	Extremely toxic substance
Briefer contact time	Potential to create an explosive atmosphere
No undesirable effects on marine organisms	Facility needs the capability to generate pure oxygen as well as an ozone generator
Decreases turbidity and odor	Use limited to filtered effluents

19.6 BROMINE CHLORIDE

Bromine chloride, a mixture of bromine and chlorine, forms hydrocarbons and hydrochloric acid upon contact with water. An excellent disinfectant, bromine chloride reacts quickly and produces no long-term residuals under normal conditions (see Table 19.5).

The chemical reactions that occur when bromine chloride contacts wastewater are similar to those that occur with chlorine disinfection, although bromamine compounds form rather than chloramines. Bromamine compounds are unstable and dissipate quickly, though they are excellent disinfectants. The compounds that form as bromamines decay are less toxic than chlorine by-products, and are not detectable in plant effluent (see Figure 19.7).

Factors that affect bromine chloride's performance as a disinfectant are also similar to chlorine. Process performance depends on effluent quality and contact time.

19.7 NO DISINFECTION

On a case-by-case basis, in very limited numbers, discharge of effluent with no disinfection is allowed. Conditions governing this decision include potential

TABLE 19.5. Bromine Chloride Advantages and Disadvantages.

Advantages	Disadvantages
Excellent disinfectant	Extremely corrosive compound with low moisture concentrations
Reacts quickly	Bromamines are unstable and dissipate quickly
Produces no long term residuals	
Bromamines decay into other less toxic compounds rapidly	

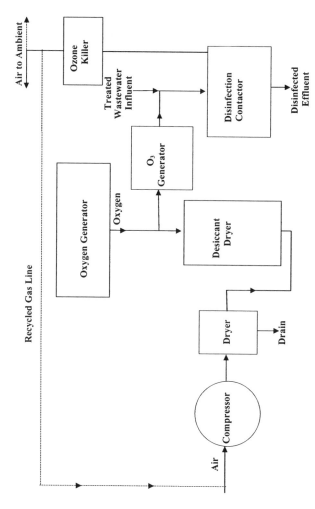

Figure 19.6 Ozone disinfection flow diagram. *Source:* Adapted from White, 1986, p. 937.

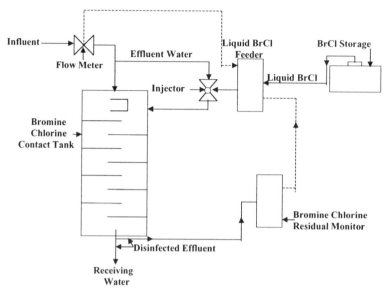

Figure 19.7 Bromine chloride feeder system.

for human contact with the discharged effluent and point of discharge, as well as many other factors.

SUMMARY

Disinfection and dechlorination are the last steps in wastewater treatment before returning the wastewater effluent to the environment. Discharge and treated wastewater reuse conclude wastewater treatment.

REFERENCE

White, G. C. *Handbook of Chlorination*, 2nd ed. New York: Van Nostrand Reinhold, 1986.

CHAPTER 20

Discharge Effluent

20.1 PROCESS PURPOSE: DISCHARGE AND REUSE

RETURNING treated wastewater to the environment safely is the chief task of effluent discharge and reuse (see Figure 20.1). This can be accomplished by a number of different methods, each of which uses water's physical behavior to accomplish wastewater's safe return to the hydrogeologic cycle. Wastewater reuse offers the attraction of both offsetting treatment costs and providing environmental benefits, especially in water-poor areas.

20.2 WASTEWATER DISCHARGE

Direct discharge into the environment is highly regulated, and meeting the NPDES standards for discharge is the goal of most wastewater treatment processes. Direct impact on the discharge environment and on the population in the discharge area are among the chief considerations for discharge. Methods to lessen effluent impact related to discharge include:

- Discharge to holding ponds or oxidation ponds allows evaporation, shifting water to vapor form and returning it to the environment.
- Discharge to ocean waters for coastal areas; this involves massively diluting the treated effluent for safe return to the environment by discharge at multiple points through a diffuser.
- Discharge through beneficial reuse allows water-scarce areas to conserve potable water supplies, and returns effluent to the environment eventually by groundwater infiltration and evapotranspiration through landscape irrigation and other methods.

221

Nonpotable water sign.

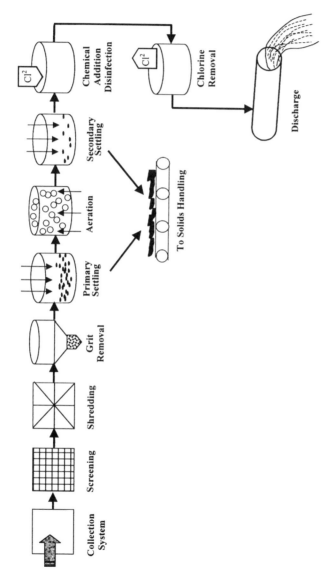

Figure 20.1 Unit processes for wastewater treatment: discharge.

Collection System

Screening

Shredding

Grit Removal

Primary Settling

Aeration

Secondary Settling

Cl² Chemical Addition Disinfection

Cl² Chlorine Removal

Discharge

To Solids Handling

223

20.3 WASTEWATER REUSE

Wastewater reuse is an idea that is quickly gaining popularity in wastewater treatment. While circumstances may dictate the necessity of wastewater reuse (in some areas, wastewater effluent is an essential water resource), now, in general, water reclamation (treatment or processing of wastewater to make it reusable) and reuse are performed with increasing regularity. With current technologies and testing for effluent quality and assurance of process safety, education for public acceptance is often a primary concern for treatment facilities. The idea that no higher quality water (unless a water surplus exists) should be used for a purpose that can tolerate a lower grade is core to acceptance of this practice.

Wastewater reuse provides many advantages, and effluent can be reused easily for several purposes. Reuse offers an alternative disposal outlet that can accept a lower level of treatment, thus protecting delicate ecosystems and offsetting costs associated with advanced treatment to meet discharge effluent regulations. Wastewater can be reused to enhance or create recreational facilities, industrial water supplies, groundwater recharge, and for direct reuse for potable water supplies (though this use is highly controversial).

20.3.1 NON-POTABLE REUSE

Current wastewater reuse practices in cities involve dual distribution systems that deliver two grades of water (one potable, the other non-potable) to the same service area. This conserves the limited high quality waters, allowing potable waters to serve a much larger population.

Using one system from a high quality potable source, and the other of reclaimed water, the reclaimed water is used for non-potable purposes. These include household, industry, commerce, and public facilities for landscape irrigation, lakes, public fountains, and environmental improvements (see Photo 20.1).

Development of reuse and reclamation projects is partially driven by the need to conserve potable water, but also by costs. Reclaiming wastewater for non-potable reuses is less expensive than meeting the nutrient removal requirements for discharge, while making use of the nutrients present in the effluent through irrigation, as well as replenishing groundwater supplies and using soil's natural filtering processes.

Reclaimed wastewater is used in applications where the product poses no health threat to consumers. These can include

- irrigation of landscaped areas (parks, athletic fields, school yards, areas around public facilities and buildings, and highway medians and shoulders) and golf courses
- irrigation of landscaped areas around family homes and nurseries

Photo 20.1 Nonpotable water sign.

- commercial activities—vehicle washing, window washing, concrete production
- use for fire protection
- toilet and urinal-flushing in industrial, commercial, and residential buildings (especially high occupancy facilities)
- boiler-feed water, make-up water for evaporative cooling towers, and irrigation of facility grounds in industries that include textile production, chemical manufacture, petroleum products, and steel manufacture

An informative publication, USEPA's *Guidelines for Water Reuse* (1992), covers the current status of practice of urban non-potable reuse. It includes all types of reclamation and reuse, and lists regulations and standards, state by state, as well as addressing international practices.

20.3.2   POTABLE REUSE

The need for reducing freshwater withdrawal by reusing and reclaiming wastewater is growing. Population increases and diminishing freshwater sources will drive this need further in the coming years. How much wastewater reuse as potable water (with plant effluent piped directly into a potable water system) will occur is difficult to judge.

Potable reuse program success hinges on one major issue: public acceptance. The reclaimed water production can be technically feasible, approved by all concerned regulatory agencies, proven safe for human consumption, and still fail because of public perception of the reclaimed water's quality.

In short, explaining to the public that they will eventually be drinking the same water they just flushed down their toilets is a hard sell. Public education is the answer to that problem, but another aspect of direct potable reuse is still not answered: health and safety concerns. The use of a pipe-to-pipe water supply of reclaimed wastewater for human consumption involves concerns over chemical and pathogen transmission, a major constraint against direct potable reuse. We do not yet know the long-term consequences of consuming water that has received such intensive chemical treatment. This is the biggest obstacle to pipe-to-pipe wastewater reuse.

SUMMARY

Wastewater effluent is not the only possible wastewater process for reuse. Wastewater treatment involves both liquid and solid wastes. The treated wastewater effluent comprises the liquid half of wastewater treatment. Water solids treatment and wastewater biosolids treatment and reuse are the topics of Part 4.

PART IV

BASICS OF WATER AND WASTEWATER SOLIDS TREATMENT AND MANAGEMENT

ID fan for incineration.

CHAPTER 21

Water Solids Management: System Overview[2]

21.1 PROCESS PURPOSE: WATER TREATMENT SLUDGES

S URFACE water influent streams include solids, which must be removed and then somehow disposed. Two principle types of sludge result from water treatment: *coagulation sludge* and *softening sludge*. After large solids are separated from the raw water by screening, suspended solids and particulate matter remain in the influent until they are removed by coagulation and flocculation. More solids wastes are formed with water softening treatment. Other sludge sources in water treatment include grit from presedimentation and solids from filter backwash, from iron and magnesium removal, and from slow sand and diatomaceous earth filtration.

The conditions of coagulant sludge are dictated by the coagulant used—generally alum (aluminum sulfate). Separated from the raw water in the sedimentation step that follows coagulation and flocculation, these water solids contain alum floc, which makes a very fine sludge that presents dewatering problems. Coagulation sludges (commonly called alum sludge) produced from clarifier operations and filter backwashing are gelatinous, and have high aluminum or iron salts concentration mixed with organic and inorganic solids. These solids are difficult to dewater, and in the past were released into water, where they caused environmental problems. Today, these wastes are processed and treated for ultimate disposal.

Softening sludges, the waste product of water softening, contain calcium carbonate and magnesium hydroxide precipitates mixed with organic and

---

[2]Source information for this chapter is from Pandit, M. and S. Das. *Sludge Disposal: Water Management Primer CE4124: Environmental Information Management.* Civil Engineering Department, Virginia Tech. http://www.ce.vt.edu/enviro2/wtprimer/sldg/sldg.html

Water solids treatment.

TABLE 21.1. Sludge Types and Concentrations.

Sludge Source and Chemical Additive	Solid Concentration before Treatment	After Thickening
Coagulation (alum)	1%	2%
Softening (lime)	1%	30%

Adapted from AWWA, 1990.

inorganic materials. Softening sludges dewater easily, and processing for disposal is standard practice (see Table 21.1).

21.2 WATER TREATMENT SLUDGE DISPOSAL REGULATIONS

While no federal regulations specifically target water treatment solids, several regulations affect use and disposal of these wastes. These include the Clean Water Act (CWA), Criteria for Classification of Solid Waste Disposal Facilities and Practices (40 CFR Part 257), the Resource Conservation and Recovery Act (RCRA), the Comprehensive Environmental Response, Compensation, and Liability Act (CERCLA), and the Clean Air Act (CAA). Direct discharges into a water course are controlled by the CWA. Any treatment facility that discharges waste into a receiving body of water must meet NPDES permit standards, in accordance with CWA Section 402. The other regulations govern other methods of disposal or use of solid wastes from water treatment. Individual states are responsible for setting state regulations that meet the federal requirements, and ensuring that the requirements are met (Burris and Smith, p. 2).

21.3 ALUM SLUDGE TREATMENT PROCESSES

The coagulant solids are separated from the water by settling, and collected for further treatment. In some locations, alum sludge is discharged into the sewerage system, where it is classified as an industrial waste, and can necessitate special processes to work it into the normal wastestream flow. In facilities that handle their own solids treatment, processes include thickening, conditioning, and dewatering before disposal.

21.4 SOFTENING SLUDGE (LIME SLUDGE) TREATMENT PROCESSES

Because of the high volume of sludge softening produces, and because these sludges form lime deposits on wastewater collection and treatment facility components, lime sludge cannot be piped into sewerage systems for disposal.

Lime sludges are thickened, dewatered, and can be centrifuged and calcified into calcium carbonate and magnesium hydroxide. This recalcined lime is a usable product—quicklime.

SUMMARY

To meet regulation requirements and ensure environmentally safe disposal, water treatment sludge must undergo treatment. We discuss the basics of treatment processes and disposal in Chapter 22.

REFERENCES

AWWA. *Water Quality and Treatment*, 1990.

Burris, B. E. and J. E. Smith. *Management of Water Treatment Plant Residuals for Small Communities*. 6th National Drinking Water and Wastewater Treatment Technology Transfer Workshop, Kansas City, MO. USEPA, 1999.

Pandit, M. and S. Das. *Sludge Disposal: Water Management Primer CE4124: Environmental Information Management*. Civil Engineering Department, Virginia Tech. http://www.ce.vt.edu/enviro2/wtprimer/sldg/sldg.html

CHAPTER 22

Water Solids Treatment and Disposal[3, 4]

22.1 WATER SLUDGE TREATMENT PROCESSES

WATER solids can be handled through a number of different processes, and a variety of disposal options are available. The process steps are thickening, coagulant recovery, conditioning, dewatering, drying, disposal and reuse, and recovered and non-recovered water handling (for more on solids equipment and handling, see Chapters 23 and 24 on wastewater biosolids management).

22.1.1 ALUM SLUDGE TREATMENT

After sedimentation, the solids are mechanically raked from the tank bottom and piped away for treatment. Typically, water treatment sludges are partially dewatered, thickened, and conditioned, then disposed of by landfill, or in some cases, land application. Because of the high aluminum content, alum sludge and cake are considered hazardous waste, and disposal must meet regulation requirements (see Photo 22.1).

Alum sludge typically has a water content before treatment of 95 to 99%. The solids content (0.1 to 0.5% inorganic clays and various organic compounds trapped in the alum floc) must be separated from the water content, often by using thickeners and settling ponds. Additional treatments with centrifuge or filters are also used. The dewatered sludge is then generally disposed of by landfill.

---

[3] Source information for this chapter is from Burris, B. E. and J. E. Smith. *Management of Water Treatment Plant Residuals for Small Communities.* 6th National Drinking Water and Wastewater Treatment Technology Transfer Workshop, Kansas City, MO. USEPA, 1999.

[4] Pandit, M. and S. Das. *Sludge Disposal: Water Management Primer CE4124: Environmental Information Management.* Civil Engineering Department, Virginia Tech. http://www.ce.vt.edu/enviro2/wtprimer/sldg/sldg.html

Water solids thickener.

Photo 22.1 Alum sludge drying beds, left side empty, right side full.

235

Alum sludge presents difficulties in dewatering. The dewatered sludge frequently can have a water content of 75%. This raises transport and landfill costs, so more effective dewatering processes are of benefit.

22.2 THICKENING

Gravity thickeners are most commonly used to concentrate water sludges, however, flotation thickeners and gravity belt thickeners are also used. Thickening directly affects conditioning and dewatering processes.

Gravity thickening: in gravity settling tanks, sludges of a specific gravity of greater than 1% are flowed at a rate that allows the solids enough time to settle. Solids thickened by gravity may also need polymers in conditioning.

Flotation thickening: this solids handling method is effective for sludges of low-density particles. More common in wastewater biosolids handling, the three types of flotation thickeners are currently attracting attention for treating water sludges. Dissolved air flotation (DAF), dispersed air flotation, and vacuum flotation all use air bubbles to absorb particles and float them to the surface for separation.

Gravity belt thickening: this emerging technology for the water industry involves direct discharge of sludge onto a horizontal porous screen. Gravity removes the water from the sludge as the sludge travels along the screen length. Polymers are generally necessary for effective solids capture with this technique.

22.3 COAGULANT RECOVERY

Acidification with sulfuric acid can be used for alum recovery; this is not a practice in common use.

22.4 CONDITIONING

Conditioning is effected by either physical or chemical treatment.

Physical conditioning: by heating the solids in a reactor, or by freezing and thawing the solids, water bound to the particles can be released, increasing solids concentrations by up to 20%.

Chemical conditioning: in chemical conditioning, the addition of ferric chloride, lime, or polymers encourages particles to give up more of the water bound with them. It is common in most mechanical thickening or dewatering processes. Raw water quality, the coagulant used, pretreatment methods, the degree of solids concentration needed, and the thickening and dewatering processes used are considerations for choosing the type of conditioner and the dosage.

22.5 DEWATERING

Water solids dewatering is typically accomplished by a variety of possible methods. These include air drying methods and mechanical methods.

Air drying: methods of sludge dewatering that remove water by evaporation, gravity, or drainage (air-drying) are easy and generally inexpensive to operate (Table 22.1). However, they also require large land areas, are affected by climate, and can be labor-intensive. Air drying methods in current use include sand drying beds, freeze-assisted sand beds, solar drying beds, vacuum assisted drying beds, and lagoons.

Mechanical dewatering: mechanical methods (Table 22.2) of sludge dewatering include belt filter presses, centrifuges, pressure filters, and vacuum filters.

22.6 DRYING

Drying can reduce transportation and disposal costs by reducing solids volume and water content. In some states, drying to a solids concentration of more than 35% before disposal is mandated by regulation. Drying can occur by either open air or mechanical means.

Open air: lagoon or solar bed methods are commonly used for drying. Evaporation mechanism is key to successful drying; some sludges can take years to reach the required concentrations.

22.7 DISPOSAL

Ultimate disposal methods possible include land application, monofill or co-disposal landfilling, direct stream discharge, sewer discharge, and residual reuse (Table 22.3). While water treatment solids do not have the fertilizer value that wastewater biosolids hold, some beneficial reuse is possible.

22.8 RECOVERABLE AND NON-RECOVERABLE WATER

The processes used to treat sludge separate the sludge into solid and liquid components. The liquid component must be either returned to the water treatment processes, as long as it will not adversely affect the main treatment processes, or to finished water quality.

SUMMARY

Biosolids systems, treatment, and disposal are covered in Chapters 23, 24, and 25.

TABLE 21.1. Air Drying Dewatering Methods.

Sand Drying Beds	Freeze-Assisted Sand Beds	Solar Drying Beds	Vacuum Assisted Drying Beds	Lagoons
Gravity drainage of free water, then decanting and evaporation. Cover beds in high rainfall areas	Freezing releases water and alters solids cellular consistency. Bound water thaws and drains before solids thaw	Paved bed on porous subbase, sand drains on edges and in the center collect and drain water	Vacuum to underside of rigid porous media pulls free water through the media and leaves even, solid cake	Used for sludge storage, thickening, dewatering and/or drying, for final disposal. 1–3 months retention time
Max loading and minimum application and removal cycles for best use	High costs for mechanical freezing, economic advantage in cold climates	Can use heavy equipment for handling and removal	Problems with poor conditioning and incomplete media cleaning, and cake removal time	Large land mass needed. Sand/solar bed modifications possible, cold climates adapt freeze methods
Effective for lime sludges, condition for alum sludges	Effective dewatering method for difficult to dewater alum sludge	Mixing and aerating sludge speeds dewatering evaporation	Conditioned sludges can dewater to 11 to 17%	Alum sludges concentrate 6–10%; lime sludge, 20–30%

Information derived from Burris and Smith, 1999, pp. 17–19.

238

TABLE 21.2. Mechanical Dewatering Methods.

Belt Filter Presses	Centrifuges	Pressure Filters	Vacuum Filters
Porous belts pass over rollers of varied diameter, applying pressure to sludge to squeeze out water	Rotational forces in a cylindrical bowl separate solids from liquids, discharging solid cake and liquid centrate separately	Fixed volume recessed plate filters: a series of plates with recessed sections filled with sludge are pressed against filter media to retain solids and pass liquid. Diaphragm filter presses: combine high pressure pumping with pressure chamber volume variation to further compress cake	Commonly used into the mid 1970s. Used to a small extent currently for water solids
Includes polymer conditioning for both lime and alum	Polymer conditioning for sludges.	Lime added to alum sludge for conditioning. No conditioning needed for lime sludge	
Effective for lime sludge (50–60% solids), alum sludge must be dewatered at low pressure (15–20% solids, certain conditions, 20–40%)	Alum dewatering rates vary according to sludge type and water source. Lime sludges dewater effectively to 55–60%	Alum sludges dewater to 30–60%, lime sludges easily dewater to 50–70% solids from a 1–3 hour cycle	Reasonably effective for lime sludges, not effective for alum sludges

Information derived from Burris and Smith, 1999, pp.19–21

TABLE 21.3. Common Disposal Techniques.

Land Application	Landfilling	Direct Stream Discharge	Sewer Discharge	Residual Reuse
Agricultural land	Co-disposal with municipal solid waste	Discharge to U.S. waters	Discharge to sewer ending at WWTP	Coagulant recovery
Forest land	Daily landfill cover extender	Discharge to water district supply canal		Lime recovery
Marginal land	Monofil WTP residuals alone	Discharge to intermittent stream bed		Low solids—landscape nursery, turf farming
Designated site	Co-disposal with WWTP solids	Discharge to dry arroyo		High solids—brick making, portland cement

Information derived from Burris and Smith, 1999, pp. 25–27.

240

REFERENCES

Burris, B. E. and J. E. Smith. *Management of Water Treatment Plant Residuals for Small Communities.* 6th National Drinking Water and Wastewater Treatment Technology Transfer Workshop, Kansas City, MO. USEPA, 1999.

Pandit, M. and S. Das. *Sludge Disposal: Water Management Primer CE4124: Environmental Information Management.* Civil Engineering Department, Virginia Tech. http://www.ce.vt.edu/enviro2/wtprimer/sldg/sldg.html

CHAPTER 23

Wastewater Biosolids Management: System Overview

23.1 PROCESS PURPOSE: WASTEWATER BIOSOLIDS TREATMENT

W ASTEWATER influent streams include solids that must be removed, and then somehow disposed of: this often ultimately involves beneficial reuse. Wastewater sludge or biosolids (the preferred term, as sludge implies a waste product and biosolids implies a substance with reuse value) must be properly treated, and biosolids management is an important part of the waste-water treatment processes.

Beneficial reuse can mean that many treatment plants use biosolids reuse as a concrete and profitable treatment method for dealing with the solids that remain after wastewater treatment.

23.2 SEWAGE BIOSOLIDS REGULATIONS

In the 1978 amendments to the Clean Water Act of 1972, the EPA's 40 CFR Part 503 Final Rules for Use and Disposal of Sewage Biosolids came into effect. The Final Rules set forth a comprehensive program for reducing the potential environmental risks and maximizing beneficial biosolids use (see Photo 23.1). In these Final Rules, the EPA

- assessed the potential for pollutants in sewage biosolids to affect public health and the environment through a number of different routes of exposure
- evaluated the risks posed by pollutants that might be present in biosolids applied to land, and considered human exposure through inhalation, direct ingestion of soil fertilized with sewage biosolids, and consumption of crops grown in the soil with sewage biosolids

243

Waste biosolids sampling point.

Photo 23.1 Sampling point required by the EPA's 40 CFR Part 503 Final Rules for Use and Disposal of Sewage Biosolids.

TABLE 23.1. Composition of Typical Raw Domestic Sewage.

Raw Domestic Sewage (99.9% Water)					
0.1% Solids					
Organic 70%			Inorganic 30%		
Proteins 65%	Carbohydrates 25%	Fats 10%	Metals	Grit	Salts

Adapted from Lester, 1992, p. 33.

- assessed the potential risk to human health through contamination of drinking water sources or surface water when biosolids are disposed of on land

23.3 PROCESS: WASTEWATER BIOSOLIDS TREATMENT

Wastewater biosolids are made up of solid particles suspended in water. Removing the water to the point that the remaining materials are easily handled is essential to biosolids treatment. This can be accomplished through drainage, thickening, or mechanical dewatering.

During wastewater treatment, the solids (sludge) are separated from the influent by settling, and collected for further treatment (Table 23.1). The collected sludge is thickened and stabilized, which improves handling, removes pathogens, and reduces the organic content. Chemical (achieved by adding lime) or aerobic/anaerobic digestion are the methods generally used in sludge stabilization. The stabilized sludge is conditioned and dewatered, then disposed of via land disposal, composting, or incineration.

23.4 PUBLIC OPINION AND ODOR CONTROL

The major public relations problem associated with composting—and with wastewater processes in general—is reportedly the production of odors. At a biosolids composting facility, sensible odor control management must take into account all the areas and components of the composting process that might generate odors. While most odor problems are generated in the composting and curing process air systems, at enclosed composting operations, odors are also generated from ancillary processes within the enclosure. Enclosed systems must control or scrub airflow within the structure prior to outside environmental release. Open areas, including biosolids handling and mixing areas, can also cause odor control problems.

Odors generally do not present a problem until the neighbors complain. Long established composting facilities may have enjoyed, at earlier times, few neighbors, and thus a lack of public awareness of their facility and the odors generated by the composting process. If the local community has expanded into, and has become a close neighbor of an existing compost facility, and if the composting process is not carefully managed (controlled), the neighbors will complain. In short, if a community perceives a compost facility in its "backyard" as a potential nuisance that will pollute the environment, decrease land values, and affect the quality of life, the struggle to bring the public onboard is not to be underestimated.

SUMMARY

Thought biosolids management is not a problem-free process, beneficial biosolids reuse does not merely present possible advantages. For many treatment plants, biosolids reuse is a concrete and profitable treatment method of dealing with that approximately .05% of wastewater influent that solids may comprise.

REFERENCES

Lester, F. N. "Sewage and Sewage Control Treatment." In R. Harrison (Ed.), *Pollution: Causes, Effects, & Control.* London: Royal Society of Chemistry, 1992.

USEPA. Standards for Use or Disposal of Sewage Sludge. Final rule, 40 CFR Part 503. *Federal Register* 58(32): 9248–9415. 19, February 1993, Washington, D.C.: U.S. Government Printing Office, 1993.

CHAPTER 24

Wastewater Biosolids Treatment

24.1 WASTEWATER BIOSOLIDS TREATMENT AND DISPOSAL ALTERNATIVES

TURNING "sludge" to "biosolids" is the goal of biosolids treatment before biosolids disposal. Alternative treatment methods and end uses are possible, and process selection depends on several factors, including the nature of the sludge to be treated, the possible end uses, and, of course, economic concerns. In ideal situations, biosolids management and disposal can have a positive effect on the municipality bottom line; in less than ideal circumstances, careful management and planning can keep costs down and provide environmentally sound alternatives to buying landfill space (for some of the alternatives available at different stages of treatment, see Table 24.1).

Turning sludge to biosolids involves thickening the sludge, reducing the sludge volume, controlling odor, killing pathogens, and removing water.

24.2 THICKENING

Biosolids thickening is a physical process accomplished by gravity (solids are allowed to settle to the bottom), flotation (solids are floated to the top), or centrifugation. Biosolids thickening removes as much water as possible before other treatment processes, and increases treatment process efficiency.

24.2.1 GRAVITY THICKENING

Similar to circular sedimentation basins in structure, gravity thickeners process thin biosolids concentrations to a more dense biosolids. Gravity thickeners

249

Biosolids cake.

TABLE 24.1. Biosolids Unit Processing Alternatives.

Thickening	Stabilization	Conditioning	Dewatering	Disposal
Gravity	Anaerobic digestion	Chemical	Vacuum	Water
Flotation	Aerobic digestion	Thermal	Filtration	Landfill
Centrifugation	Thermal		Pressure filtration	Soil conditioning
	Chemical		Centrifugation	Composting
				Incineration

are usually used to handle watery excess biosolids from the activated biosolids process, and the process is common in plants where the biosolids are sent directly to digesters instead of to primary tanks. Gravity thickening can also concentrate biosolids from primary tanks, or a mixture of primary and excess biosolids, prior to high rate digestion.

Gravity thickening tanks generally house slowly moving biosolids scrapers with an attached vertical picket-fence-like structure. This moving structure agitates the biosolids and dislodges entrapped liquid and gas bubbles (McGhee, 1991). Biosolids are pumped continuously from the settling tank to the thickener, which has a low overflow rate. The excess water overflows and solids concentrate on the bottom. This method can produce biosolids with a solids content of 10% or more from an original solids concentration of 2% (see Figure 24.1).

24.2.2 FLOTATION THICKENING

In flotation thickening, a tiny air bubble attaches to suspended solid particles, causing the solids to separate from the water and float. With the air bubble, the solid particles have a specific gravity lower than water's.

Dissolved air flotation depends on forming small diameter bubbles under pressure. Current flotation practice uses two general approaches to pressurization:

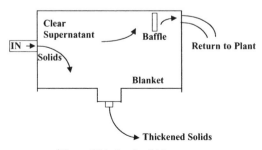

Figure 24.1 Gravity thickener.

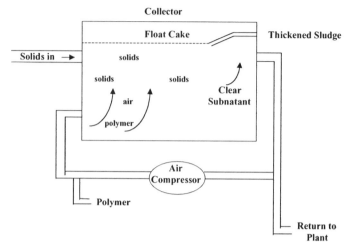

Figure 24.2 Flotation thickener.

(1) air charging and pressurization of recycled clarified effluent or some other flow used for dilution, with subsequent addition to the feed biosolids; and (2) air charging and pressurization of the combined dilution liquid and feed biosolids (see Figure 24.2 and Photo 24.1).

Variables that affect flotation thickening systems include the type and quality of the biosolids used, pressure, feed solids concentration, recycle ratio, detention time, air-to-solids ratio, solids and hydraulic loading rates, and chemicals.

Flotation thickening works best for activated biosolids. Equal or greater concentrations may be achieved by combining biosolids in gravity thickening units.

24.2.3 CENTRIFUGATION

Primarily used in dewatering (see Section 24.6.2.3), centrifugation can also thicken a variety of biosolids, though for thickening purposes, it is generally limited to processing waste activated biosolids (Metcalf & Eddy, 1991). Centrifuges are compact, simple, flexible, self-contained units with relatively low capital costs, high maintenance and power costs, and poor solids-capture efficiency without chemical use.

24.3 STABILIZATION

Stabilizing biosolids reduces the volume of the thickened biosolids further, eliminates offensive odors, reduces the possibility of putrefaction, and renders the remaining solids relatively pathogen-free (Peavy et al., 1991). Biosolids

Photo 24.1 DAF thickening biosolids sampling point.

stabilization can be accomplished by anaerobic digestion, aerobic digestion, and thermal and chemical methods.

24.3.1 ANAEROBIC DIGESTION

Anaerobic digestion biosolids digestion is carried out in the absence of free oxygen by anaerobic organisms—in short, anaerobic decomposition. The solid matter in raw biosolids is about 70% organic and 30% inorganic. Much of the water in wastewater biosolids is "bound" water, unable to separate from the solids. Facultative and anaerobic organisms break down the molecular structure of these solids and set free the "bound" water.

Anaerobic digestion can be simplified to two steps: (1) conversion of organic materials to volatile acid wastes, and (2) conversion of volatile acids into methane (Haller, 1995). In the first step (waste conversion), acid-forming bacteria attack soluble or dissolved complex solids (fats, proteins, and carbohydrates), forming organic acids and gases, such as carbon dioxide and hydrogen sulfide, in a rapid process called *acid fermentation*. In the *acid digestion* process that follows, acid-forming bacteria attack the organic acids and nitrogenous compounds, liquefying (creating supernatant liquid) at a much slower rate (Masters, 1991).

In the second stage (the period of intensive digestion, stabilization, and gasification), the more resistant nitrogenous materials (the proteins, amino-acids, and others) are attacked by methane-forming microorganisms that produce large volumes of gases with a high percentage of methane (CH_4) and carbon dioxide. The relatively stable remaining solids are only slowly putrescible, and can be disposed of without creating objectionable conditions. They have value in agriculture for liquid land application and compost.

Anaerobic digesters are airtight cylindrical, rectangular, or egg-shaped tanks in which digestion occurs. Standard-rate, high-rate, and two-stage digesters are common. The most inefficient and unstable digesters, standard-rate digesters generally do not heat or mix the sludge. High-rate digesters heat the sludge to optimum temperature and increase digestion rates by mixing. Two-stage digesters separate digested solids from the supernatant liquid, sometimes achieving high-rate digestion levels.

24.3.2 AEROBIC DIGESTION

An extension of the activated biosolids aeration process, in aerobic digestion (Table 24.2), waste primary and secondary biosolids are continually aerated for long periods. During this extended aeration, the microorganisms enter a phase (the endogenous stage) where materials previously stored by the cell oxidize, reducing the biologically degradable organic matter (see Photo 24.2).

TABLE 24.2. Aerobic Digestion System.

Advantages	Disadvantages
Produces a humus-like, biologically stable end product	High power costs
Produces an end product with no odors, thus simple land disposal is feasible.	Requires a supply of oxygen that is energy-consumptive
Can be accomplished with low capital costs compared to anaerobic and other digestion methods	Does not always settle well in subsequent thickening processes
Produces sludge, with good dewatering characteristics, that drains and re-dries well in sand drying beds	Does not dewater easily by vacuum filtration
Produces volatile solids reduction levels equal to anaerobic digestion	Variable solids reduction efficiency with varying temperature changes
Produces supernatant liquors with a lower BOD (generally lower than 100 ppm) than anaerobic digestion	
Fewer operating problems than with anaerobic methods, requiring less skilled labor for facility operation	
Recovers more of the biosolids basic fertilizer values than anaerobic digestion	

Adapted from Metcalf & Eddy, 1991.

In simple terms, in the endogenous stage, food supplies to microbial life are depleted to the point where the microorganisms begin to consume their own protoplasm, oxidizing it to carbon dioxide, water, and ammonia. As the digestion process continues, the ammonia is further converted to nitrates. Eventually, the oxygen uptake rate levels off and the biosolids matter is reduced to inorganic matter and relatively stable volatile solids.

Aerobic digestion is affected by biosolids temperature, rate of biosolids oxidation, biosolids loading rate, system oxygen requirements, biosolids age, and biosolids solids characteristics.

24.3.3 THERMAL STABILIZATION

In thermal stabilization, the bound water in the biosolids is released by heating the biosolids in a pressurized tank for short periods of time. Thermal treatment is used both for stabilization and for biosolids conditioning.

Exposing the biosolids to heat and pressure coagulates the solids, breaks down the cell structure, and reduces both hydration and the hydrophilic (water-loving) nature of the solids. Liquids can then be separated from the biosolids by decanting and pressing.

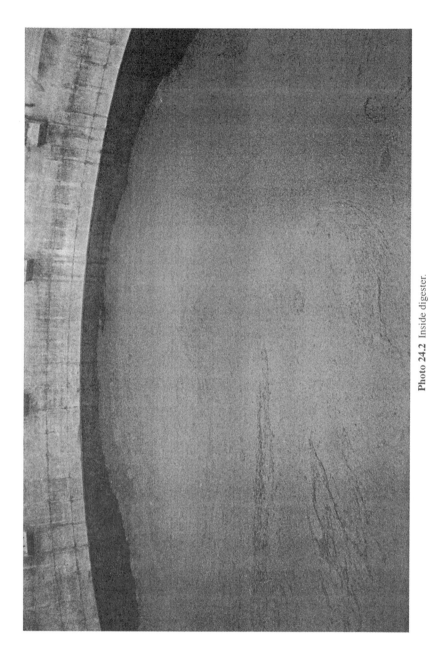

Photo 24.2 Inside digester.

256

24.4 CHEMICAL APPLICATION

In chemical stabilization, the biosolids mass is treated with chemicals to stabilize the solids. Chemical costs can be high (in some cases cost-prohibitive).

24.4.1 CHLORINE STABILIZATION

The chemical reactions that occur in chlorine stabilization happen almost instantaneously. A high dose of chlorine gas applied directly to the biosolids in an enclosed reactor produces stabilized biosolids. Chlorine stabilization produces some breakdown of organic material, and forms carbon dioxide and nitrogen, with little volatile solids reduction.

Total solids, suspended solids, and volatile solids concentrations are similar to that of the raw biosolids. Chlorine-treated biosolids have a low pH value and require pH adjustment before conditioning. They may also contain undissolved heavy metals and chlorinated compounds, limiting their suitability for land application. For these reasons, chlorine stabilization is not often used for biosolids stabilization.

24.4.2 LIME STABILIZATION

Lime stabilization (successfully used for many years) can be used to treat raw primary, waste activated, septage, and anaerobically digested biosolids. Lime works by increasing the pH to levels high enough to destroy most microorganisms, and to limit odor production. Lime stabilization denatures organic matter, it does not destroy it, so the treated sludge must be disposed of before it putrefies.

The process involves mixing a large enough quantity of lime with the biosolids to increase and sustain the pH of the mixture to 12 or more. Along with reducing bacterial hazards and odor to negligible levels, lime stabilization also improves vacuum filter performance; provides a satisfactory means of stabilizing biosolids prior to ultimate disposal; and by maintaining the pH above 12 for 2 hours or more, substantially improves the total reduction in microorganisms over that obtained in digestion processes (McGhee, 1991).

24.5 CONDITIONING

In biosolids conditioning, the solids are treated with chemicals or by various other means to improve production rate, cake solids content, and solids capture, all of which prepare the biosolids for dewatering processes. Several different biosolids conditioning processes are available; the most commonly used are chemical addition and heat treatment.

24.5.1 CHEMICAL CONDITIONING

Chemical conditioning (biosolids conditioning) prepares the biosolids for better and more economical treatment with dewatering equipment by reducing the biosolids moisture content to from 60 to 85%. Chemicals used for conditioning include organic polymers, alum, ferrous sulfate, and ferric chloride, with or without lime, and others, all of which are more easily applied in liquid form (see Photo 24.3) (Metcalf & Eddy, 1991). The choice of chemical type to use depends on the nature of the biosolids to be conditioned, and local costs.

Adding chemicals to the biosolids lowers or raises its pH value to the point that small particles coagulate into larger ones, and the biosolids give up water more readily. pH values vary with the chemical used, and different biosolids (primary, secondary, and digested biosolids), and even different biosolids of the same type, have different optimum pH values (for more information on chemical feeders, see Section 14.2.5.1).

24.5.2 THERMAL CONDITIONING

Thermal conditioning destroys the biological cells in biosolids, which "permits a degree of moisture release not achieved in other conditioning processes" (McGhee, 1991, p. 497). The two most common basic processes for thermal treatment in current use are wet air oxidation and heat treatment.

In wet air oxidation, biosolids are flamelessly oxidized in a reaction vessel at temperatures between 450 and 550°F, and at pressures of about 1200 psig. Heat treatment is similar to wet air oxidation, however, heat treatment temperatures fall between 350 and 400°F, with pressures of 150 to 300 psig (Table 24.3).

Both these processes release water that is bound up in the biosolids, facilitating the dewatering process (Davis and Cornwell, 1991). Wet air oxidation reduces the biosolids to an ash, and heat treatment improves the dewaterability of the biosolids. Heat treatment is more widely used than the oxidation process.

24.6 DEWATERING

According to the USEPA (1982), the objectives of dewatering "are to remove water and thereby reduce the [biosolids] volume, to produce a [biosolids] which behaves as a solid and not a liquid, and to reduce the cost of subsequent treatment and disposal processes" (p. 2).

Dewatering also

- reduces need for space, fuel, labor, equipment, and size of the composting facility
- affects the amount of bulking agent needed for composting
- lowers costs for transporting biosolids

Photo 24.3 Polymer storage tank.

TABLE 24.3. Thermal Conditioning: Wet Air Oxidation.

Advantages	Disadvantages
Does not require preliminary dewatering or drying Produces a more readily dewaterable biosolids than with chemical conditioning Provides effective biosolids disinfection	Oxidized ash must be separated from the water in an additional process Heat treatment ruptures the cell walls of microbial organisms, releasing bound organic material, and creating the added problem of treating a highly polluted liquid sidestream from the cells High initial capital costs Higher levels of competence and training of operating personnel Significant production of odorous gases

Biosolids dewatering can be accomplished in a variety of ways. When the necessary land space is available for small quantities of sludge, lagoons and drying beds are a good option. For large quantities of sludge, a variety of mechanical methods are common. These mechanical methods include: vacuum filtration, pressure filtration, and centrifuging (Qasim, 1999, pp. 728–734).

24.6.1 NATURAL DEWATERING METHODS

Sludge drying beds and lagoons are a low cost dewatering option, commonly used in small to medium sized systems where land-use issues are not a factor. The oldest technique for dewatering, new methods have improved drying bed efficiency, and new methods are making this option more attractive for large plants as well. Sand beds, paved beds, wire-wedge and vacuum assisted drying beds are now available. In general, drying beds offer drying times of constant and rapid drainage for the more sophisticated wire-wedge and vacuum assisted drying beds, to 30 to 40 days for conventional drying beds. Solids capture ranges from 60 to 70% for sand drying beds to 90 to 100% capture for paved beds.

Lagoons are similar to drying beds in that the dewatered biosolids are removed periodically, and the lagoon reused. Climate and sludge depth affect evaporation time. In general 3 to 6 months is needed to reach a solids level of 20 to 40%. Solids capture can achieve 90 to 100% (see Photos 24.4a to 24.4d).

24.6.2 MECHANICAL DEWATERING METHODS

Mechanical methods offer advantages of speed in processing, but are generally more costly in terms of equipment.

Photo 24.4a Sludge drying bed.

Photo 24.4b Sludge drying bed (drying in progress).

Photo 24.4c Sludge drying bed (drying in progress).

263

Photo 24.4d Sludge drying bed (process finished).

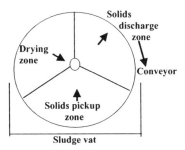

Figure 24.3 Vacuum filter.

24.6.2.1 Vacuum Filtration

Vacuum filtration for biosolids dewatering involves laying filtering media (a cloth of cotton, wool, nylon, fiberglass, a plastic or stainless steel mesh, or a double layer of stainless steel coil springs) over a drum. The drum (on a horizontal axis) is submerged to about one-fourth its depth in a tank of conditioned biosolids. As a portion of the drum rotates slowly in and out of the biosolids, an arrangement of valves and piping function to apply a vacuum to the inner side of the filter medium. This draws water out of the biosolids and holds the biosolids against the media surface. The vacuum's pressure continues while the drum rotates, carrying the biosolids into the atmosphere, pulling water away from the biosolids. A moist mat or cake remains on the drum's outer surface, which is scraped, blown, or lifted away from the drum just before the drum rotates again into the biosolids tank (see Figure 24.3).

24.6.2.2 Pressure Filtration

Two types of pressure filtration systems are common: belt filter (see Figure 24.4) or recessed plate filter presses. In either pressure system, the filtration

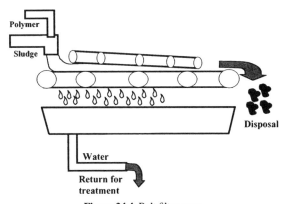

Figure 24.4 Belt filter press.

TABLE 24.4. Pressure Filtration System.

Advantages	Disadvantages
Can be accomplished quickly in a small space Good performance (> 30% cake solids in some cases) when compared to the vacuum filter (18% cake solids produced in some cases)	Batch operation High operation and maintenance costs

method resembles vacuum filtration. While recessed plate filter press units are most common, belt filter press usage has increased (Table 24.4).

As in vacuum filtration, recessed plate filter presses use a porous media to separate biosolids from water. Captured solids in the media pores build up on the media surface. Biosolids pumps provide the energy to force the water through the media.

24.6.2.3 Centrifugation

In centrifugation, solids separate from liquid through sedimentation and centrifugal force. In a typical unit, the biosolids mass feeds via a stationary feed tube along the centerline of the bowl through a screw conveyor hub. The screw conveyor, mounted inside the rotating conical bowl, rotates at a slightly lower speed than the bowl. As biosolids leave the end of the feed tube, they are accelerated, pass through ports in the conveyor shaft, and distribute to the bowl periphery. The solids settle through the liquid, compact by centrifugal force against the walls of the bowl, and the screw conveyor carries them to the bowl's drying (or beach) area, an inclined section where further dewatering occurs before the solids discharge. The separated liquid discharges continuously over adjustable weirs at the bowl's opposite end (see Figure 24.5 and Photo 24.5).

SUMMARY

The thickened, stabilized, conditioned and dewatered remains of what was once sludge are now ready for one of several disposal methods. These alternatives are covered in Chapter 25.

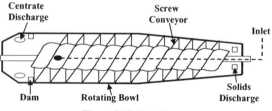

Figure 24.5 Centrifuge.

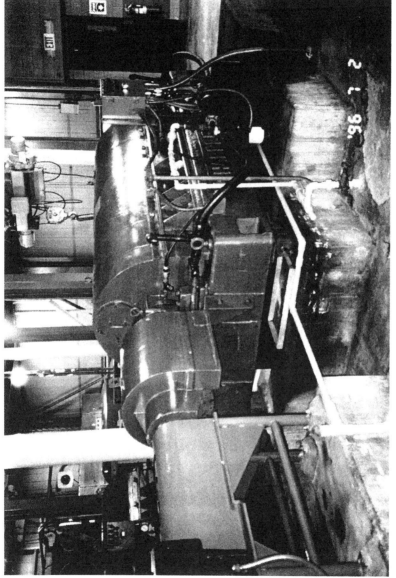

Photo 24.5 Centrifuge.

REFERENCES

Davis, M. L. and D. A. Cornwell. *Introduction to Environmental Engineering*, 2nd ed. New York: McGraw-Hill, Inc., 1991.

Haller, E. J. *Simplified Wastewater Treatment Plant Operations.* Lancaster, PA: Technomic Publishing Co., Inc., 1995.

Masters, G. M. *Introduction to Environmental Engineering & Science.* Englewood Cliffs, NJ: Prentice Hall, 1991.

McGhee, T. J. *Water Supply and Sewerage.* New York: McGraw-Hill, Inc., 1991.

Metcalf & Eddy. *Wastewater Engineering: Treatment, Disposal, & Reuse*, 3rd ed. New York: McGraw-Hill, Inc., 1991.

Peavy, H. S., D. R. Rowe, and G. Tchobanoglous. *Environmental Engineering.* New York: McGraw-Hill, Inc., 1991.

Qasim, S. R. *Wastewater Treatment Plants: Planning, Design, and Operation*, 2nd ed. Lancaster, PA: Technomic Publishing Co., Inc., 1999.

USEPA. Dewatering Municipal Wastewater Sludges. EPA-625/1-82-014. Cincinnati: Center for Environment Research Information, 1982.

CHAPTER 25

Wastewater Biosolids Disposal

25.1 DISPOSAL ALTERNATIVES

WITH or without biosolids thickening, stabilization, conditioning, and/or dewatering, wastewater treatment facilities must have a plan or routine to follow in disposing of treated or untreated biosolids (see Figure 25.1 and Table 25.1).

Biosolids produced during the wastewater treatment process may contain concentrated levels of contaminants originally contained in the wastewater. Any disposal method must take these contaminants into account. In ultimate disposal, the goal must not be to merely shift the original pollutants in the wastestream to a final disposal site where they may become free to contaminate the environment. Effective biosolids disposal involves economically and environmentally sound disposal or reuse that adheres to the EPA's 40 CFR Part 503 Final Rules for Use and Disposal of Sewage Biosolids.

25.2 DISPOSAL METHODS

Methods for the disposal of biosolids include:

- disposal in water (no longer permitted)
- disposal on land
- reuse as a fertilizer or soil conditioner
- reuse in land reclamation projects
- reuse in composting.
- Incineration and ash disposal

Incinerator.

TABLE 25.1. Biosolids Management Options.

Use/Disposal Practice	Mass of Sewage Biosolids Used/ Disposed by POTW Size			
	>100 mgd	10–100 mgd	1–10 mgd	<1 mgd
Landfill	518.4	673.6	495.4	110.4
Land application	387.7	664.7	538.1	178.1
Surface disposal	79.5	264.6	122.1	87.2
Incineration	382.8	346.3	124.7	10.5

Source: USEPA, 1993, pp. 9248–9415.

25.2.1 WATER

Water or ocean disposal was once thought an economical method of biosolids disposal, used by communities in coastal areas or along major rivers. It is now illegal. Common practice for biosolids disposal (raw or digested) was to either pipe it directly offshore, or pump it to barges that carried it to deep water, distant enough from the shore so that dilution would (hopefully) prevent shoreline problems.

As sewage biosolids quantities increased, so did pollutional loads. When ocean-dumped biosolids began to creep back toward the shore, affecting beaches along the upper East Coast of the U.S., the obvious problems inherent in ocean disposal caused widespread concern. The dumping of sewage biosolids was prohibited by Congress in 1992.

25.3 LAND

Interest in biosolids disposal on land increased when ocean dumping was prohibited, and when new air quality regulations began to impact biosolids

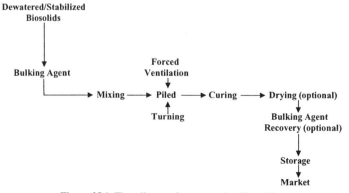

Figure 25.1 Flow diagram for composting biosolids.

incineration. Common (and closely regulated) methods of land disposal include:

- landfilling
- application as fertilizer or soil conditioner
- land reclamation

25.3.1 LANDFILLING

Once widely used, landfilling with biosolids is rapidly becoming a past practice, the result of two major influences:

- Large densely populated areas are running out of space in landfills for any type of waste.
- EPA 503 regulations directly affect disposal methods for wastewater biosolids.

Biosolids fill is confined almost entirely to well digested biosolids with no applicable amount of raw or undigested mass. These digested biosolids can be exposed to air without creating serious or widespread odor nuisances.

With three common disposal methods no longer always feasible because of regulation and/or costs, beneficial reuse through land disposal became a much more attractive possibility.

25.3.2 LAND APPLICATION: SOIL CONDITIONING/FERTILIZING

Land application of biosolids is popular because it is relatively simple to accomplish at a relatively reasonable cost (Table 25.2).

On estimate, the fertilizer nitrogen requirements of about 0.35 million hectares (2.5 acre = 1 hectare) of cropland could be supplied with sewage biosolids (Laws, 1993). In short, all the sewage biosolids the United States produces could provide only a small fraction of the annual fertilizer nitrogen required for U.S. crop production.

25.3.3 LAND RECLAMATION

Biosolids land application's most dramatic success has been strip-mined land restoration (Laws, 1993). Establishing vegetation on such land is extremely difficult, because typically, strip-mined land lacks nutrients and organic matter, has low pH, poor water retention, and often contains high levels of toxic metals in the soil (Sopper and Kerr, 1981).

In 1993, an estimated 15 to 20% of municipal sewage biosolids was applied to land (Laws, 1993). Biosolids applied to restore lands devastated through

TABLE 25.2. Soil Conditioning and Fertilizing with Land Applied Biosolids.

Advantages	Disadvantages
Can contain as much nitrogen and phosphorus as farmyard manure, is agriculturally valuable Recycles nutrients, including nitrogen, phosphorous, and potassium Provides traces of other nutrients considered more or less indispensable for plant growth, including calcium, copper, iron, magnesium, manganese, sulfur, and zinc Relatively simple to accomplish at a relatively reasonable cost Reduced disposal costs Humus-like quality furnishes plant food, and increases water holding capacity and tillage, improving heavy soils for seed bed use Helps to reduce soil erosion	Nutrient content of sewage biosolids varies greatly and is always lower in potassium than farmyard manure Often contains non-essential metallic elements in quantities possibly toxic to plants and animals Can contain heavy metals and other toxic agents. Applied too often and for too long, it increases in the levels of heavy metals, including copper, cobalt, boron, lead, mercury, and others. These toxins are not easily removed

strip-mining and other activities is an ecologically beneficial practice, and also a very practical disposal method.

25.4 COMPOSTING

Composting, a process that converts organic wastes into a soil amendment, has been a process known to be useful since ancient times (see Photo 25.1). When organic wastes are partially decomposed by bacteria, worms, and other living organisms, the result is a valuable fertilizer and soil conditioner. Composted materials were and are used to prevent erosion, provide nutrients to the soil, and replenish depleted organic matter lost through farming (Corbitt, 1990).

With several alternate biosolids disposal methods outlawed, unwanted, and/or declared unsafe, composting has become an attractive option for waste disposal. Today, with the use of treatment methodologies that include heat-drying, treatment with alkaline materials, and composting, municipalities convert biosolids into useful products safe for unrestricted public use.

Biosolids composting has four goals:

- to stabilize the product
- to control odor

Photo 25.1 Compost pile.

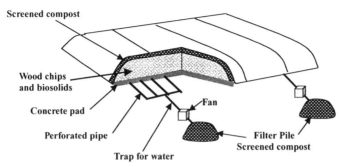

Figure 25.2 Aerated static pile.

- to dry the material enough to handle
- to raise the biosolids temperature high enough to kill pathogens

25.4.1 AERATED STATIC PILE COMPOSTING

In the static aerated pile method of composting, the homogenized mixture of bulking agent (coarse hardwood wood chips) and dewatered biosolids is piled by front-end loaders onto a large concrete composting pad. There, the mixture is mechanically aerated via PVC plastic pipe embedded within the concrete slab. In a 26-day "active" composting period, adequate air and oxygen supports aerobic biological activity in the compost mass, and reduces the compost mixture's heat and moisture content (see Figure 25.2).

Aeration is an important process control parameter in the aerated static pile composting system.

- It supplies oxygen for biological degradation of organic solids in the biosolids and wood chips (aerated static pile or ASP model).
- It removes heat generated by the biological activity in the compost pile, and excess moisture from the compost mix.
- Improperly aerated composting leads to the onset of anaerobic conditions and putrefactive odors.

25.4.2 CURING AND DRYING

After 26 days, the compost is dried or cured, or a combination of both.

In curing, compost stabilizes as microorganisms metabolize the nutrients that remain in the mixture. A 30-day curing period is required before final sampling/testing and distribution. Generally considered an extension of aeration processes, curing and any subsequent storage can create elevated temperatures, although somewhat lower than average temperatures associated with initial composting. Curing ensures total odor dissipation and total pathogen destruction. The cured compost product should not have an unpleasant odor.

Drying (accomplished by drawing or blowing air through aeration pipes in troughs, by mechanical mixing, or using a combination of these methods) is optional, but is usually necessary if compost is to be recycled as a bulking agent, or if screening is required. Drying can occur in a roofed drying shed or structure to protect the compost from inclement weather, or if weather conditions permit, on any hard surface area.

25.4.3 SCREENING

Specialized screening equipment separates compost from the wood chips after curing and/or drying, recycling the wood chips for reuse. The compost moves to distribution for marketing in bulk or bags.

The most common compost screening devices are shaker screens and trommel screens.

25.4.3.1 Shaker Screens

Shaker screens move and shake up and down, sifting the compost through the screen and separating the wood chips to another location. Screen size depends on bulking agent size.

The major disadvantage of shaker screens is their tendency to clog, and the extensive amount of maintenance required to keep them on-line.

25.4.3.2 Trommel Screens

Trommel screens (long cylindrical rotating screens) are usually placed on an angle so that materials flow through them. Trommel drums rotate on wheels and generally can be tilted from 3 to 12 degrees. Large brushes mounted on top of the trommel drum extend through the screen to prevent clogging, making the trommel unit "self-cleaning."

In the trommel drum, compost materials and bulking agents separate by a tumbling action. The smaller compost materials fall through the grate while the bulking agent eventually discharges at one end. Screened compost collects on an underbelt conveyor system that travels the length of the trommel drums. Conveyors transport both wood chips and compost from the screens (see Figure 25.3).

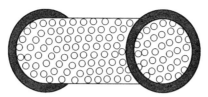

Figure 25.3 End view of trommel screen. Adapted from Spellman, F. R. *Wastewater Biosolids to Compost.* Technomic Publishing Co., Inc., 1997, p. 157.

25.5 INCINERATION

Biosolids incineration (the complete destruction of biosolids by heat) is not actually a means of disposal, but of volume reduction to ash, a reuse product (a resource) with some value. Incinerating biosolids is not only a beneficial process; it produces a beneficial reuse product: *biosolids ash.*

25.5.1 INCINERATION ADVANTAGES

Compared to several other biosolids management options (especially land-filling or lagoons), biosolids incineration presents some obvious advantages:

- reduces biosolids volume and weight
- provides immediate reduction with long-term residence times
- avoids transportation costs (on-site incineration)
- controlled air discharges and abiding by regulatory requirements maintains air quality
- leaves sterile and thus usually harmless ash residue
- requires relatively small disposal area compared to land requirements for lagoons or land burial (Brunner, 1984)

25.5.2 INCINERATION PROCESS

Biosolids incineration can occur in four steps:

(1) Biosolids temperature is raised to 212°F.

(2) Water evaporates from the biosolids.

(3) Increase water vapor temperature and air temperature.

(4) Biosolids temperature elevated to the ignition point of the volatiles (USEPA, 1978, p. 5).

A key factor in incineration is the solids content of the biosolids. Before incineration, the biosolids must be dewatered (often by centrifugation), because even though the heat value of biosolids is relatively high, excessive water content means auxiliary fuel must be used to maintain incinerator combustion (USEPA, 1978). Providing auxiliary fuel impacts incinerator operation economics. Increased solids content directly affects the net heat value of the feedstock, allowing autogenous (self-sustained) combustion and eliminating auxiliary fuel costs.

25.6 TYPES OF INCINERATORS

Four different types of incinerators are currently commonly used for biosolids incineration: the multiple hearth furnace (MHF), the fluid bed furnace (FBF), the cyclonic furnace (single rotary furnace), and the electric furnace (EF) (Table 25.3). While the multiple hearth furnace is most commonly used for biosolids incineration, most new installations are fluid-bed furnaces. Cyclonic and electric furnaces are not commonly used in the United States for biosolids incineration (see Photo 25.2).

25.6.1 MULTIPLE HEARTH FURNACE (MHF)

The multiple hearth furnace (MHF) provides easy operation, and good capacity for handling wide fluctuations in feed loading rate and biosolids cake of

TABLE 25.3. Furnace Type Disadvantages.

Multiple Hearth	Fluid Bed	Cyclonic	Electric
Extensive maintenance and preventive maintenance on rabble arm assemblies, hearth refractories, and fuel and air systems	Problems with feed and temperature control	Not widely used	Not widely used
Several hours to cool to ambient temperature	Mechanized machinery (screw feeds and pumps) can jam with dry biosolids		In some locations, cost of electricity can be very high
Heating furnace too quickly damages refractories	Sand scaling affects venturi scrubber system		Large floor space requirement
	Serious scrubber erosion can occur from excessive bed material carryover		

Photo 25.2 Sludge entering furnance.

279

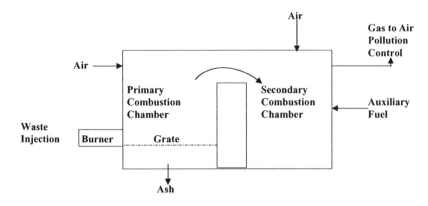

2-Stage Hearth Incinerator

Figure 25.4 Hearth incinerator.

differing quality. Vertically oriented and cylindrically shaped (see Figure 25.4 and Photo 25.3), multiple hearth furnaces are refractory lined with a steel shell, and contain several (from 4 to 14) horizontal refractory hearths, one above the other.

In multiple hearth furnaces, a central shaft (which contains a fan-fed inner tube called the cold air tube) runs the height of the furnace. The central shaft supports rabble arms (toothed scrapers or plows) above each hearth, with either two or four rabble arms per hearth. The rabble arms rotate and rake the biosolids spirally across the hearth (see Photo 25.4). Each rabble arm connects to the cold air tube, and has a return tube to send the heated cooling air to the space between the cold air tube and the central shaft's shell (an exhaust passageway for the cooling air). Returned (heated) cooling air enters the lowest hearth again as preheated combustion air.

In a typical MHF operation, biosolids fed at the periphery of the top hearth are raked toward the center, then drop to the hearth below. The second hearth drop holes are at the periphery of the bed, and biosolids are raked outward to drop to the next hearth. Alternating drop hole locations on each hearth, and counter-current airflow or rising exhaust gases and descending biosolids provide good contact between the hot combustion gases and the biosolids feed to ensure complete combustion.

Usually divided into four distinct zones, multiple hearth furnaces provide a series of combustion chambers (Table 25.4). In the first zone (*drying zone*), most of the remaining water evaporates. The *zone of combustion* (at the central hearths) burns the biosolids at temperatures that range from 1,400 to 1,700°F. In the *fixed carbon burning zone*, carbon oxidizes to carbon dioxide, and in the *cooling zone*, re-injecting heat to the incoming combustion air cools the ash residue. Although the sequence of these zones is always the same, the number

Photo 25.3 Multi-hearth incinerator.

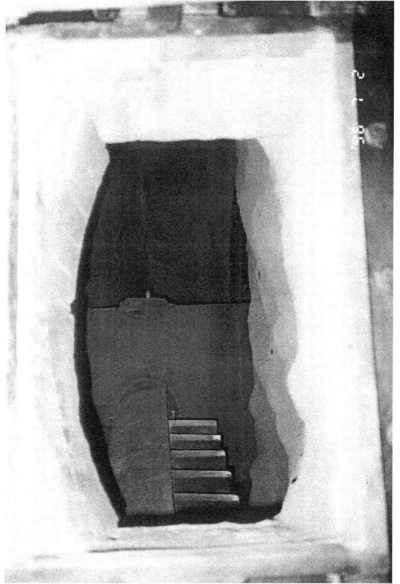

Photo 25.4 Incinerator rabble arms with teeth.

TABLE 25.4. **Multiple Hearth Furnace Operations.**

Zones	Operation
Drying zone	Dewatered solids supplied at top hearth outer edge
	Rotating rabble arms move the solids slowly to the hearth center
	Solids dry by hot gases produced by burning on lower hearths
	Dry solids pass to the lower hearths (may take several hearths)
	Process repeats, but this time from inside edge to outside edge
Combustion zone	High temperature on the lower hearth ignites the solids
	Burned solids pass to the lower hearths (may take several hearths)
	Burning continues to completion
Fixed carbon burning zone	Carbon oxidizes to carbon dioxide
	Ash passes to the lower cooling hearths
Cooling zone	Ash temperatures decrease from withdrawing heated air
	Ash is discharged for disposal or reuse
	Internal equipment is continuously cooled by air flowing inside center column and rabble arms

of hearths in each zone depends on feedstock quality, operational conditions, and furnace design (USEPA, 1978).

25.6.2 FLUID-BED FURNACE (FBF)

Fluid-bed furnaces are vertically oriented, cylindrically shaped, refractory-lined steel shells (reactors) that contain a sand bed (media) and fluidizing air diffusers (orifices). They range in size from 9 to 25 ft diameters (see Figure 25.5 for cross section). At rest, the sand bed is approximately 2.5 ft thick. It rests on a refractory-lined grid or brick dome that provides orifices (truyere) through which air is injected into the bed at pressure to fluidize the bed, expanding it by approximately 80 to 100%. Sand bed temperature is kept between 1,400 and 1,500°F by auxiliary burners. Some installations include heat control by water spray or a heat removal system.

In fluid-bed furnaces (Table 25.5), ash is carried to the furnace top and removed by air pollution control devices. Sand carried out with the ash must be replaced. Biosolids enter the furnace either above or directly into the sand bed.

Fluid bed furnace excess air requirements reduce supplemental fuel needs and reduce heat losses from heating and exhausting excess air. Complete contact

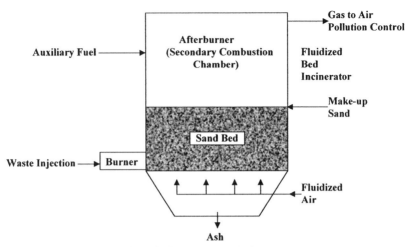

Figure 25.5 Fluidized bed incinerator.

between biosolids solids and the combustion gases occurs through mixing caused by airflow through the bed, and direct sand bed biosolids injection. The high amount of retained heat in the sand means the system can operate as little as four hours per day with little or no reheating.

25.6.3 CYCLONIC FURNACE

Cyclonic furnaces (single rotary furnaces) are vertically oriented, single-rotary hearth furnaces with cylindrically shaped, refractory-lined domed steel shells (see Figure 25.6). Cyclonic furnaces use one rotating hearth and a fixed rabble arm (scraper). A screw-type feeder deposits the biosolids near the

TABLE 25.5. Fluid Bed Furnace Operations.

	Operation
1	Air is pumped into the bottom of the unit
2	Air flow expands (fluidizes) the sand bed
3	Fluidized sand bed heats to between 1,200 and 1,500°F
4	Auxiliary fuel added when needed to maintain temperature
5	Solids injected into the heated sand bed
6	Moisture immediately evaporates
7	Organic matter ignites and reduces to ash
8	Residues grind to fine ash by sand movement
9	Fine ash particles flow up and out with exhaust gases
10	Air pollution control processes remove ash particles
11	Exhaust gas stack oxygen analyzers control air flow rate

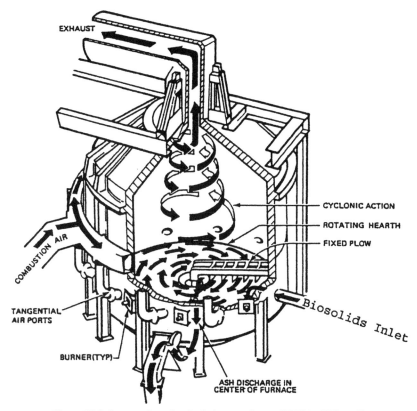

Figure 25.6 Cross section of cyclonic furnace. *Source:* USEPA, 1978, p. 13.

periphery of the rotating hearth, and the rabble arm moves biosolids materials from the outer edge of the hearth to the center, where they discharge as ash.

Cyclonic furnaces have the advantage of low capital costs and low fuel requirements. A competitive alternative furnace for biosolids incineration, the cyclonic furnace's high exhaust temperatures allows for easier compliance with air quality standards than the other furnace types.

25.6.4 ELECTRIC FURNACE

Electric or infrared (radiant heat) furnaces operate using horizontal conveyors. A steel alloy woven wire belt passes through a rectangular refractory-lined chamber equipped with electrical radiant heating elements (see Figure 25.7). Furnaces range from $4' \times 20'$ to $9.5' \times 100'$ in length.

Electric furnaces are usually divided into feed, drying, combustion, and ash discharge zones. Because electric furnaces can provide complete burn of

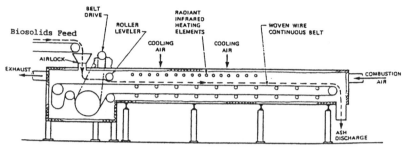

Figure 25.7 Cross section of an electric furnace. *Source:* USEPA, 1978, p. 15.

the biosolids without agitation, they generate very low levels of particulate emissions.

25.7 ASH DISPOSAL VERSUS REUSE

Wastewater biosolids and biosolids ash are waste products with some value, and should be reused, not simply disposed of. For safe biosolids ash reuse, it must be processed in accordance with EPA's 503 Rule to ensure negligible risk to human health and the environment (see Photo 25.5).

In the past, biosolids ash had been reused in limited applications (Table 25.6). The USEPA's 503 Rule, and increasingly more stringent regulations governing disposal at state and local levels, has raised interest in developing additional beneficial biosolids ash applications. These applications include:

- supplementary material added as bulking agents for biosolids composting
- additives for construction materials (concrete aggregates for road building or other construction activities)
- landfill cover material

TABLE 25.6. Ash Disposition.

Application	% of Use (Rounded)
Land application	42
Co-disposal	20
Incineration	13.5
Surface disposal	9
Distribution and marketing	6
Ocean disposal	5
Monofilling	2
Other	2.5

Source: USEPA, 1990.

Photo 25.5 Biosolids ash storage bin.

- composite material used for novelty product production
- composite material used for shoreline erosion control product production

SUMMARY

Biosolids reuse, as well as wastewater reclamation and reuse, are essential parts of the technological processes for ensuring that our water supply continues to meet our demands, as well as our needs. With a finite supply of water and a growing population, a constant cycle of use, treatment, and reuse becomes more and more necessary for providing for our water needs, in all aspects of water and wastewater processing.

REFERENCES

Brunner, C. R. *Incinerator Systems: Selection and Design.* New York: Van Nostrand Reinhold Company, Inc., 1984.

Corbitt, R. A. *Standard Handbook of Environmental Engineering.* New York: McGraw-Hill, Inc., 1990.

Laws, E. A. *Aquatic Pollution,* 2nd ed. New York: John Wiley & Sons, Inc., 1993.

Sopper, W. E. and S. N. Kerr. *Revegetating Strip-Mined Land with Municipal Sewage Sludge.* Project Summary USEPA Report 600/52-81-182. Washington, D.C. Government Printing Office, 1981.

USEPA. *Operations Manual: Sludge Handling & Conditioning,* 1978.

USEPA. *National Sewage Sludge Survey,* 1990.

USEPA. *Standards for Use or Disposal of Sewage Sludge,* 1993.

Summary of Major Amendment Provisions for the 1996 SDWA Regulations

Definition	• *Constructed conveyances*, such as cement ditches, used primarily to supply substandard drinking water to farm workers are now SDWA protected.
Contaminant Regulation	• Old contamination selection requirement (EPA regulates 25 new contaminants every 3 years) deleted.
	• EPA must evaluate at least 5 contaminants for regulation every 5 years, addressing the most risky first, and considering vulnerable populations.
	• EPA must issue *Cryptosporidium* rule (enhanced surface water treatment rule) and disinfection by product rules under agreed deadlines. The Senate provision giving industry veto power over EPA's expediting the rules was deleted.
	• EPA is authorized to address "urgent threats to health" using an expedited, streamlined process.
	• No earlier than 3 years after enactment, no later than the date EPA adopts the State II DBP rule, EPA must adopt a rule requiring disinfection of certain *groundwater* systems, and provide guidance on determining which systems must disinfect. USEPA may use cost-benefit provisions to establish this regulation.
Risk Assessment, Management, and Communication	• Requires cost/benefit analysis, risk assessment, vulnerable population impact assessment, and development of public information materials for EPA rules.

289

Standard Setting

- Standard setting provision allows, but doesn't require, EPA to use risk assessment and cost/benefit analysis in setting standards.
- Cuts back Senate's process from 3 to 2 steps to issue standards, deleting requirement of Advanced Notice of Proposed Rule Making.
- Risks to vulnerable populations must be considered.
- Has cost/benefit and risk/risk as discretionary USEPA authority. "Sound Science" provision is limited to standard setting and scientific decisions.
- Standard reevaluated every 6 years instead of every 3 years as current law.

Treatment Technologies for Small Systems

- Establishes new guidelines for EPA to identify best treatment technology for meeting specific regulations.
- For each new regulation, USEPA must identify affordable treatment technologies that achieve compliance for 3 categories of small systems: those serving 3,301–10,000, those serving 501–3,000 and those serving 500 or fewer.
- For all contaminants other than microbials and their indicators, the technologies can include package systems as well as point-of-use and point-of-entry units owned and maintained by water systems.
- EPA has 2 years to list such technologies for current regulations, and 1 year to list such technologies for the surface water treatment rule.
- EPA must identify best treatment technologies for the same system categories for use under variances. Such technologies do not have to achieve compliance, but must achieve maximum reduction, be affordable, and protect public health.
- EPA has 2 years to identify variance technologies for current regulations.

Limited Alternative to Filtration

- Allows systems with fully controlled pristine watersheds to avoid filtration if EPA and State agree health is protected through other effective inactivation of microbial contaminants.
- EPA has 4 years to regulate recycling of filter backwash.

Effective Date of Rules	• Extends compliance time from 18 months (current law) to 3 years, with available extensions of up to 5 years total.
Arsenic, Sulfate, Radon	• *Arsenic*: requires EPA to set new standard by 2001 using new standard setting language, after more research and consultation with the NAS (National Academy of Sciences). The law authorizes $2.5 million/year for 4 years for research.
	• *Sulfate*: EPA has 30 months to complete a joint study with the Federal Centers for Disease Control (CDC) to establish a reliable dose-response relationship. Must consider sulfate for regulation within 5 years. If EPA decides to regulate sulfate, it must include public notice requirements and allow alternative supplies to be provided to at-risk populations.
	• *Radon*: requires EPA to withdraw its proposed radon standard and to set a new standard in 4 years, after NAS conducts a risk assessment and a study of risk-reduction benefits associated with various mitigation measures. Authorizes cost/benefit analysis for radon, taking into account costs and benefits of indoor air radon control measures. States or water systems obtaining EPA approval of a multimedia radon program in accordance with EPA guidelines would only have to comply with a weaker "alternative Maximum Contaminant Level" for radon, that would be based on the contribution of outdoor radon, to indoor air.
State Primacy	• Primacy states have 2 years to adopt new or revised regulations, no less stringent than federal ones, and allows 2 or more if EPA finds necessary and justified.
	• Provides states with interim enforcement authority between the time they submit their regulations to EPA and EPA approval.
Enforcement and Judicial Review	• Streamlines EPA administrative enforcement, increases civil penalties, clarifies enforceability of lead ban and other previously ambiguous requirements, allows enforcement to be suspended in some cases to encourage system consolidation or restructuring, requires states to have administrative penalty authority, and clarifies provisions for judicial review of final EPA actions.

Public Right to
Know

"Consumer Confidence Reports" provision requires
consumers be told at least annually:

- the levels of regulated contaminants detected in tap
  water

- what the enforceable maximum contaminant levels
  and the health goals are for the contaminants (and
  what those levels mean)

- the levels found of unregulated contaminants required
  to be monitored

- information on the system's compliance with health
  standards and other requirements

- information on the health effects of regulated
  contaminants found at levels above enforceable
  standards, and on health effects of up to 3 regulated
  contaminants found at levels below EPA enforceable
  health standards where health concerns may still
  exist

- EPA's toll-free hotline for further information

- Governors can waive the requirement to mail these
  reports for systems serving under 10,000 people, but
  systems must still publish the report in the paper.

- Systems serving 500 or fewer people need only
  prepare the report and tell their customers it's
  available.

- States can later modify the content and form of the
  reporting requirements.

- The public information provision modestly improves
  public notice requirements for violations (such as
  requiring "prominent" newspaper publication instead
  of buried classified ads). States and USEPA must
  prepare annual reports summarizing violations.

Variances and
Exemptions

- Provisions for small system variances make minor
  changes to current provisions regarding exemption
  criteria and schedules.

- States are authorized to grant variances to systems
  serving 3,300 or fewer people but need EPA approval
  to grant variances to systems serving between 3,301
  and 10,000 people. Such variances are available only
  if EPA identifies an applicable variance technology
  and systems install it.

- Variances only granted to systems that cannot afford to comply (as defined by state criteria that meet EPA guidelines) through treatment, alternative sources or restructuring and when states determine that the terms of the variance ensure adequate health protection. Systems granted such variances have 3 years to comply with its terms and may be granted an extra 2 years if necessary, and states must review eligibility of such variances every 5 years thereafter.

- Variances not allowed for regulations adopted prior to 1986 for microbial contaminants or their indicators.

- EPA has 2 years to adopt regulations specifying procedures for granting or denying such variances, and for informing consumers of proposed variances and pertinent public hearings. They also must describe proper operation of variance technologies and eligibility criteria. USEPA and the Federal Rural Utilities Service have 18 months to provide guidance to help states define affordability criteria.

- EPA must periodically review state small system variance programs and may object to proposed variances and overturn issued variances if objections are not addressed. Also, customers of a system for which a variance is proposed can petition USEPA to object.

- New York may extend deadlines for certain small, unfiltered systems in 9 counties to comply with federal filtration requirements.

Capacity Development

- States must acquire authority to ensure that community and nontransient-noncommunity systems beginning operation after 10/1/1999 have technical, managerial and financial capacity to comply with SDWA regulations. States that fail to acquire authority lose 20% of their annual state revolving loan fund grants.

- States have a year to send USEPA a list of systems with a history of significant noncompliance and 5 years to report on the success of enforcement mechanisms and initial capacity development efforts. State primacy agencies must also provide progress reports to governors and the public.

- States have 4 years to implement strategy to help systems acquire and maintain capacity before losing portions of their SRLF grants.

- USEPA must review existing capacity programs and publish information within 18 months to help states and water systems implement such programs. USEPA has 2 years to provide guidance for ensuring capacity of new systems, and must describe likely effects of each new regulation on capacity.

- The law authorizes $26 million over 7 years for grants to establish small water systems technology assistance centers to provide training and technical assistance. The law also authorizes $1.5 million/year through 2003 for USEPA to establish programs to provide technical assistance aimed at helping small systems achieve and maintain compliance.

Operator
Certification

- Requires all operators of community and nontransient-noncommunity systems to be certified. EPA has 30 months to provide guidance specifying minimum standards for certifying water system operators, and states must implement a certification program within 2 years or lose 20% of the SRLF grants.

- States with such programs can continue to use them as long as EPA determines they are substantially equivalent to its program guidelines.

- EPA must reimburse states for the cost of certification training for operators of systems serving 3,300 or fewer people, and the law authorizes $30 million/year through 2003 for such assistance grants.

State Supervision
Program

- Authorizes $100 million/year through 2003 for public water system supervision grants to states.

- Allows EPA to reserve a state's grant should EPA assume primacy and, if needed, use SRLF resources to cover any shortfalls in PWSS appropriations.

Drinking Water
Research

- EPA is authorized to conduct drinking water and groundwater research, and is required to develop a strategic research plan, and to review the quality of all such research.

Water Return Flows	• Repeals the provision in current law that allows businesses to withdraw water from a public water system (such as for industrial cooling purposes), then to return the used water—perhaps with contamination—to the water system's pipe.
Enforcement	• Expands and clarifies EPA's enforcement authority in primacy and nonprimacy states, and provides for public hearings regarding civil penalties ranging from $5,000–$25,000.
	• Provides enforcement relief to systems that submit a plan to address problems by consolidating facilities or management, or transferring ownership.
	• States must obtain authority to issue administrative penalties, which cannot be less than $1,000/day for systems serving over 10,000 people.
	• EPA can assess civil penalties as high as $15,000/day under its emergency powers authority.

Glossary

Absorption any process by which one substance penetrates the interior of another substance.

Activated Sludge the solids formed when microorganisms are used to treat wastewater using the activated sludge treatment process: mixing primary effluent with bacteria-laden sludge, which is then agitated and aerated to promote biological treatment. This process speeds breakdown of organic matter in raw sewage undergoing secondary treatment. Activated sludge includes organisms, accumulated food materials, and waste products from the aerobic decomposition process.

Adsorption the process by which one substance is attracted to and adheres to the surface of another substance without actually penetrating its internal structure.

Advanced Waste Treatment a treatment technology used to produce an extremely high quality discharge.

Aeration a physical treatment method that promotes biological degradation of organic matter. The process may be passive (when waste is exposed to air) or active (when a mixing or bubbling device introduces air).

Aerobic conditions in which free, elemental oxygen is present. Also used to describe organisms, biological activity, or treatment processes that require free oxygen.

Aerobic Bacteria a type of bacteria that requires free oxygen to carry out metabolic function.

Alum Cake dewatered alum sludge.

Alum Sludge solids removed from sedimentation of raw water that has undergone coagulation, flocculation and sedimentation.

Aluminum Sulfate [Al$_2$(SO$_4$)$_3$] (alum)   coagulant added to raw water to form floc for solids removal in water treatment.

Anaerobic   conditions in which no oxygen (free or combined) is available. Also used to describe organisms, biological activity, or treatment processes that function in the absence of oxygen.

Anoxic   conditions in which no free, elemental oxygen is present, and the only source of oxygen is combined oxygen such as that found in nitrate compounds. Also used to describe biological activity or treatment processes that function only in the presence of combined oxygen.

Autogenous/Autothermic Combustion   (incinerator) the burning of a wet organic material where the moisture content is at such a level that the heat of combustion of the organic material is sufficient to vaporize the water and maintain combustion. No auxiliary fuel is required except for start-up.

Average Monthly Discharge Limitation   the highest allowable discharge over a calendar month.

Average Weekly Discharge Limitation   the highest allowable discharge over a calendar week.

Beneficial Uses   the many ways water can be used, either directly by people, or for their overall benefit.

Biochemical Oxygen Demand (BOD)   the amount of oxygen required by bacteria to stabilize decomposable organic matter under aerobic conditions.

Biochemical Oxygen Demand (BOD$_5$)   the amount of organic matter that can be biologically oxidized under controlled conditions (5 days @ 20°C in the dark).

Biological Treatment   a process that uses living organisms to bring about chemical changes.

Biosolids   From *Merriam-Webster's Collegiate Dictionary, Tenth Edition* (1998): biosolid *n* (1977)—solid organic matter recovered from a sewage treatment process and used especially as fertilizer—usually used in plural.

Biosolids Cake   the solid discharged from a dewatering apparatus.

Biosolids Concentration   the weight of solids per unit weight of biosolids.

Biosolids Moisture Content   the weight of water in a biosolids sample divided by the total weight of the sample. Normally determined by drying a biosolids sample and weighing the remaining solids, the total weight of the biosolids sample equals the weight of water plus the weight of the dry solids.

Biosolids Quality Parameters   three main EPA parameters used in gauging biosolids quality: (1) the relevant presence or absence of pathogenic organisms, (2) pollutants, and (3) the degree of attractiveness of the biosolids to vectors.

Bucketing   simple, effective, but labor intensive method for cleaning large amounts of debris from a sewer line. Workers load buckets from within the line, hauling the solids to the surface for disposal.

Buffer   a substance or solution that resists changes in pH.

Building Service   collection system connection and pipe that carries wastewater flow from the generation point to a main.

Cake Solids Discharge Rate   the dry solids cake discharged from a centrifuge.

Carbonaceous Biochemical Oxygen Demand (CBOD$_5$)   the amount of biochemical oxygen demand that can be attributed to carbonaceous material.

Centrate   the effluent or liquid portion of a biosolids removed by or discharged from a centrifuge.

Chemical Oxygen Demand (COD)   the amount of chemically oxidizable materials present in the wastewater.

Chemical Treatment   a process that results in the formation of a new substance or substances. The most common chemical water/wastewater treatments include coagulation, disinfection, water softening, and oxidation.

Chlorine   a strong oxidizing agent that has strong disinfecting capability. A yellow-green gas, it is extremely corrosive, and is toxic to humans in extremely low concentrations.

Chlorine Demand   a measure of the amount of chlorine that will combine with impurities and therefore will not be available to act as a disinfectant.

Clarifier   a device designed to permit solids to settle or rise and be separated from the flow. Also known as a settling tank or sedimentation basin.

Clean Water Act (CWA)   federal law dating to 1972 (with several amendments) with the objective to restore and maintain the chemical, physical, and biological integrity of the nation's waters. Its long-range goal is to eliminate the discharge of pollutants into navigable waters, and to make and keep national waters fishable and swimmable.

Clean Zone   any part of a stream upstream of the point of pollution entry.

Cleanout Points   collection system points that allow access for cleaning equipment and maintenance into the sewer system.

Coagulants   chemicals that cause small particles to stick together to form larger particles.

Coagulation   a chemical water treatment method that causes very small suspended particles to attract one another and form larger particles. This is accomplished by adding a coagulant that neutralizes the electrostatic charges on the particles that cause them to repel each other. The larger particles are easier to trap, filter, and remove.

Coliform   a type of bacteria used to indicate possible human or animal contamination of water.

Collectors or Subcollectors   collection system pipes that carry wastewater flow to trunk lines.

Combined Sewer   a collection system that carries both wastewater and storm water flows.

Comminution   a process used to shred solids into smaller, less harmful particles.

Community Water System   a public water system that serves at least 15 service connections used by year-round residents, or regularly serves at least 25 year-round residents.

Composite Sample   a combination of individual samples taken in proportion to flow.

Contact Time   the length of time the disinfecting agent and the wastewater remain in contact.

Contaminant   a toxic material found as a residue in or on a substance where it is not wanted.

Cross Connection   any connection between safe drinking water and a non-potable water or fluid.

$C \times T$ Value   the product of the residual disinfectant concentration C, in milligrams per liter, and the corresponding disinfectant contact time T, in minutes. Minimum $C \times T$ values are specified by the Surface Water Treatment Rule as a means of ensuring adequate kill or inactivation of pathogenic microorganisms in water.

Daily Discharge   the discharge of a pollutant measured during a calendar day or any 24-hour period that reasonably represents a calendar day for the purposes of sampling. Limitations expressed as weight are total mass (weight) discharged over the day. Limitations expressed in other units are average measurements of the day.

Daily Maximum Discharge   the highest allowable values for a daily discharge.

Delayed Inflow   stormwater that may require several days or more to drain through the sewer system. This category can include the discharge of sump pumps from cellar drainage, as well as the slowed entry of surface water through manholes in ponded areas.

Demand   the chemical reactions that must be satisfied before a residual or excess chemical will appear.

Detention Time   the theoretical time water remains in a tank at a given flow rate.

Dewatering   the removal or separation of a portion of water present in a sludge or slurry.

Direct Flow   those types of inflow that have a direct stormwater runoff connection to the sanitary sewer and cause an almost immediate increase in wastewater flows. Possible sources are roof leaders, yard and areaway drains, manhole covers, cross connections from storm drains and catch basins, and combined sewers.

Direct Potable Reuse the piped connection of water recovered from wastewater to a potable water-supply distribution system or a water treatment plant, often implying the blending of reclaimed wastewater.

Direct Reuse the use of reclaimed wastewater that has been transported from a wastewater reclamation point to the water reuse site, without intervening discharge to a natural body of water (e.g., agricultural and landscape irrigation).

Discharge Monitoring Report (DMR) the monthly report required by the treatment plant's NPDES discharge permit.

Disinfect to inactivate virtually all recognized pathogenic microorganisms but not necessarily all microbial life (cf. pasteurize or sterilize).

Disinfectant (1) any oxidant, including but not limited to chlorine, chlorine dioxide, chloramine, and ozone added to water in any part of the treatment or distribution process, that is intended to kill or inactivate pathogenic microorganisms. (2) A chemical or physical process that kills pathogenic organisms in water. Chlorine is often used to disinfect sewage treatment effluent, water supplies, wells, and swimming pools.

Disinfectant Contact Time (T in CT calculation) the time (T) in minutes that it takes for water to move from the point of disinfectant application or the previous point of disinfection residual measurement to a point before or at the point where residual disinfectant concentration (C) is measured. Where only one C is measured, T is the time in minutes that it takes for water to move from the point of disinfectant application to a point before or at where residual disinfectant concentration (C) is measured.

Disinfection the addition of a substance (for example, chlorine, ozone, or hydrogen peroxide) that destroys or inactivates harmful microorganisms or inhibits their activity. Also the selective destruction of disease-causing organisms. All the organisms are not destroyed during the process. This differentiates disinfection from sterilization, which is the destruction of all organisms.

Disinfection By-Products compounds formed by the reaction of a disinfectant such as chlorine with organic material in the water supply.

Dissolved Oxygen (DO) free or elemental oxygen that is dissolved in water.

Dose the amount of chemical being added in milligrams/liter.

Drinking Water Standards water quality standards measured in terms of suspended solids, unpleasant taste, and microbes harmful to human health. Drinking water standards are included in state water quality rules.

Drinking Water Supply　any raw or finished water source that is or may be used as a public water system, or as drinking water by one or more individuals.

Drying Hearth　a solid surface in an incinerator on which wet waste materials (or waste matter that may turn to liquid before burning) are placed to dry, or to burn with the help of hot combustion gases.

Effluent　the flow leaving a tank, channel, or treatment process.

Effluent Limitations　standards developed by the EPA to define the levels of pollutants that can be discharged into surface waters, or any restriction imposed by the regulatory agency on quantities, discharge rates, or concentrations of pollutants that are discharged from point sources into state waters.

Electrodialysis　water using ion-selective membranes and an electric field to separate anions and cations in solution.

Estuaries　coastal bodies of water that are partly enclosed.

Evaporation　the process by which water as liquid changes to water vapor.

Facultative　organisms that can survive and function in the presence or absence of free, elemental oxygen.

Facultative Bacteria　a type of anaerobic bacteria that can metabolize its food either aerobically or anaerobically.

Fecal Coliform　a type of bacteria found in the bodily discharges of warm-blooded animals. Used as an indicator organism.

Federal Water Pollution Control Act (1972)　under the Act, the objective "to restore and maintain the chemical, physical, and biological integrity of the nation's waters" is outlined. This 1972 Act and subsequent Clean Water Act amendments are the most far-reaching water pollution control legislation ever enacted.

Feed Rate　the amount of chemical being added in pounds per day.

Filtrate　the effluent or liquid portion of a biosolids removed by, or discharged from, a centrifuge.

Filtration　a physical treatment method for removing solid (particulate) matter from water by passing the water through porous media such as sand or a man-made filter.

Flashpoint　the lowest temperature at which evaporation of a substance produces sufficient vapor to form an ignitable mixture with air, near the surface of the liquid.

Floc　solids that join together to form larger particles that will settle better.

Flocculation　the water treatment process following coagulation that uses gentle stirring to bring suspended particles together to form larger, more settleable clumps called floc.

Flume   a flow rate measurement device.

Flushing   line clearing technique that adds large volumes of water to the sewer at low pressures to move debris through the collection system.

Food-to-Microorganism Ratio (F/M)   an activated sludge process control calculation based upon the amount of food (BOD_5 or COD) available per pound of mixed liquor volatile suspended solids.

Friction Head   the energy needed to overcome friction in the piping system. It is expressed in terms of the added system head required.

Grab Sample   an individual sample collected at a randomly selected time.

Grit   heavy inorganic solids such as sand, gravel, egg shells, or metal filings.

Groundwater   the freshwater found under the earth's surface, usually in aquifers. Groundwater is a major source of drinking water, and concern is growing over areas where leaching agricultural or industrial pollutants or substances from leaking underground storage tanks are contaminating groundwater.

Head   the equivalent distance water must be lifted to move from the supply tank or inlet to the discharge. Head can be divided into three components: *static head, friction head*, and *velocity head.*

Horizontal Directional Drilling (HDD)   technique used to drill or bore a tunnel through the soil and pull or push new pipe in behind the drill head.

Hydraulic Line-Cleaning Devices   devices for clearing built-up debris or blockages from sewer lines. These devices include hydraulic tools such as balls, kites, pills, pigs, scooters, and bags. They work by partially plugging a flooded upstream main. The movement of the tool itself and the force of the water pressure from the partially blocked line work to loosen and flush away debris.

Hydrologic Cycle   literally, the water-earth cycle: the movement of water in all three of its physical forms—water, vapor, and ice—through the various environmental mediums (air, water, biota, and soil).

Hygroscopic   a substance that readily absorbs moisture.

Incineration   an engineered process using controlled flame combustion to thermally degrade waste material.

Indirect Potable Reuse   the potable reuse by incorporation of reclaimed wastewater into a raw water supply; the wastewater effluent is discharged to the water source, mixed, and assimilated with it, with the intent of reusing the water instead of as a means of disposal. This type of potable reuse is becoming more common as water resources become less plentiful.

Indirect Reuse   the use of wastewater reclaimed indirectly by passing it through a natural body of water, or the use of groundwater that has been recharged with reclaimed wastewater. This type of potable reuse commonly occurs whenever an upstream water user discharges wastewater effluent into a watercourse that serves as a water supply for a downstream user.

Industrial Wastewater   wastes associated with industrial manufacturing processes.

Infiltration   water entering the collection system through cracks, joints, or breaks. Infiltration includes steady inflow, direct flow, total inflow, and delayed inflow.

Influent   water, wastewater, or other liquid flowing into a reservoir, basin, or treatment plant.

Inorganic   mineral materials such as salt, ferric chloride, iron, sand, gravel, etc.

Interceptors   collection system pipes that carry wastewater flow to the treatment plant.

Jetting   line cleaning technique that cleans and flushes the line in a single operation, using a high pressure hose and a variety of nozzles to combine the advantages of hydraulic cleaning with mechanical cleaning.

Junction Boxes   collection system constructions that occur when individual lines meet and are connected.

Land Application   discharge of wastewater onto the ground for treatment or reuse.

Lift Stations   pump installations designed to pump wastes to a higher point through a force main, when gravity flow does not supply enough force to move the wastewater through the collection system.

Lime Sludge   solids removed from water softening processes.

Line-Cleaning   see *Hydraulic Line-Cleaning* and *Mechanical Line-Cleaning.*

Mains   collection system pipes that carry wastewater flow to collection sewers.

Manholes   collection system entry points that allow access into the sewerage system for inspection, preventive maintenance, and repair.

Maximum Contaminant Level (MCL)   the maximum allowable concentration of a contaminant in drinking water, as established by state and/or federal regulations. Primary MCLs are health related and mandatory. Secondary MCLs are related to the water quality aesthetic considerations and are highly recommended, but not required.

Mean Cell Residence Time (MCRT)   the average length of time a mixed liquor suspended solids particle remains in the activated sludge process. May also be known as sludge retention time.

Mechanical Line-Cleaning   methods such as rodding or bucketing used to clean stoppages and blockages from sewer lines.

mg/L   an expression of the weight of one substance contained within another. Commonly used to express weight of a substance within a given weight of water and wastewater, it is sometimes expressed as parts per million (ppm), which is equal to mg/L.

Milligrams/Liter (mg/L)   a measure of concentration equivalent to parts per million (ppm) [see (*mg/L*)].

Mixed Liquor   the combination of return activated sludge and wastewater in the aeration tank.

Mixed Liquor Suspended Solids (MLSS)   the suspended solids concentration of the mixed liquor.

Mixed Liquor Volatile Suspended Solids (MLVSS)   the concentration of organic matter in the mixed liquor suspended solids.

Moisture Content   the amount of water per unit weight of biosolids. The moisture content is expressed as a percentage of the total weight of the wet biosolids. This parameter is equal to 100 minus the percent solids concentration.

National Pollutant Discharge Elimination System (NPDES)   a requirement of the CWA that discharges meet certain requirements prior to discharging waste to any water body. It sets the highest permissible effluent limits, by permit, prior to making any discharge.

Near Coastal Water Initiative   initiative developed in 1985 to provide for management of specific problems not dealt with in other programs for waters near coastlines.

Nitrogenous Oxygen Demand (NOD)   a measure of the amount of oxygen required to biologically oxidize nitrogen compounds under specified conditions of time and temperature.

Nonbiodegradable   a substance that does not break down easily in the environment.

NPDES Permit   the National Pollutant Discharge Elimination System permit that authorizes the discharge of treated wastes and specifies the condition that must be met for discharge.

Nutrients   substances required to support living organisms. Usually refers to nitrogen, phosphorus, iron, and other trace metals.

Organic   materials that consist of carbon, hydrogen, oxygen, sulfur, and nitrogen. Many organics are biologically degradable. All organic compounds can be converted to carbon dioxide and water when subjected to high temperatures.

Osmosis   the natural tendency of water to migrate through semipermeable membranes from the weaker solution to the more concentrated solution, until hydrostatic pressure equalizes the chemical balance.

Oxidation   when a substance gains oxygen, or loses hydrogen or electrons in a chemical reaction. One of the chemical treatment methods.

Oxidizer   a substance that oxidizes another substance.

Part Per Million   an alternative (but numerically equivalent) unit used in chemistry is milligrams per liter (mg/L).

Pathogenic   disease causing. A pathogenic organism is capable of causing illness.

Physical Treatment   any process that does not produce a new substance (e.g., in wastewater treatment, screening, adsorption, aeration, sedimentation, and filtration).

Pipe Bursting   trenchless technology method that destroys the old pipe while pulling the new pipe in behind.

Planned Reuse   the deliberate direct or indirect use of reclaimed wastewater without relinquishing control over the water during its delivery.

Point Source   any discernible, defined, and discrete conveyance from which pollutants are or may be discharged.

Pollutant   any substance introduced into the environment that adversely affects the usefulness of the resource.

Pollution   the presence of matter or energy whose nature, location, or quantity produces undesired environmental effects. Under the Clean Water Act, for example, the term is defined as a man-made or man-induced alteration of the physical, biological, and radiological integrity of water.

Potable Water Reuse   a direct or indirect augmentation of drinking water with reclaimed wastewater that is highly treated to protect public health.

Precipitation   atmospheric water that falls to earth as rain (a liquid) or snow, sleet, or hail (solids).

Pressure   the force exerted per square unit of surface area. May be expressed as pounds per square inch.

Pretreatment   any physical, chemical, or mechanical process used before the main water/wastewater treatment processes. It can include screening, presedimentation, and chemical addition. Also the practice of industry removing toxic pollutants from their wastewaters before they are discharged into a municipal wastewater treatment plant.

Primary Disinfection   the initial killing of *Giardia* cysts, bacteria, and viruses.

Primary Drinking Water Standards   regulations on drinking water quality (under SDWA) that are considered essential for preservation of public health.

Primary Treatment   the first step of treatment at a municipal wastewater treatment plant. It typically involves screening and sedimentation to remove materials that float or settle.

Publicly Owned Treatment Works (POTW)   a waste treatment works owned by a state, local government unit, or Indian tribe, usually designed to treat domestic wastewaters.

Receiving Waters a river, lake, ocean, stream, or other water source into which wastewater or treated effluent is discharged.

Recharge the process by which water is added to a zone of saturation, usually by percolation through the soil.

Reclaimed Wastewater wastewater that, as a result of wastewater reclamation, is suitable for a direct beneficial use or a controlled use that would not otherwise occur.

Recovery Zone the point in a stream where, as the organic wastes decompose, the stream quality begins to return to more normal levels.

Residual the amount of disinfecting chemical remaining after the demand has been satisfied.

Return Activated Sludge Solids (RASS) the concentration of suspended solids in the sludge flow being returned from the settling tank to the head of the aeration tank.

Reverse Osmosis (RO) solutions of differing ion concentration are separated by a semipermeable membrane. Typically, water flows from the chamber with lesser ion concentration into the chamber with the greater ion concentration, resulting in hydrostatic or osmotic pressure. In RO, enough external pressure is applied to overcome this hydrostatic pressure, thus reversing the flow of water. This results in the water on the other side of the membrane becoming depleted of ions and demineralized.

Rodders mechanical method used to clear obstructions like heavy root accumulations or large soft obstructions from collection system lines to restore flow.

Safe Drinking Water Act (SDWA) federal law passed in 1974 to establish federal standards for drinking water quality, protect underground sources of water, and set up a system of state and federal cooperation to assure compliance with the law.

Sanitary Sewer collection system that carries human wastes in wastewater from residences, businesses, and some industry to the treatment facility.

Sanitary Wastewater wastes discharged from residences and from commercial, institutional, and similar facilities that include both sewage and industrial wastes.

Screening a pretreatment method that uses coarse screens to remove large debris from the water to prevent clogging of pipes or channels to the treatment plant.

Scum the mixture of floatable solids and water that is removed from the surface of the settling tank.

Secondary Disinfection the maintenance of a disinfectant residual to prevent regrowth of microorganisms in the water distribution system.

Secondary Drinking Water Standards regulations developed under the Safe Drinking Water Act that established maximum levels of substances affecting the aesthetic characteristics (taste, color, or odor) of drinking water.

Secondary Treatment the second step of treatment at a municipal wastewater treatment plant. It uses growing numbers of microorganisms to digest organic matter and reduce the amount of organic waste.

Sedimentation physical treatment method that reduces the velocity of water in basins so the suspended material settles out by gravity.

Septic wastewater with no dissolved oxygen present. Generally characterized by black color and rotten egg (hydrogen sulfide) odor.

Septic Zone the point in a stream where pollution causes dissolved oxygen levels to sharply drop, affecting stream biota.

Settleability a process control test used to evaluate the settling characteristics of the activated sludge. Readings taken at 30 to 60 minutes are used to calculate the settled sludge volume (SSV) and the sludge volume index (SVI).

Settled Sludge Volume the volume in percent occupied by an activated sludge sample after 30 to 60 minutes of settling. Normally written as SSV with a subscript to indicate the time of the reading used for calculation (SSV_{60} or SSV_{30}).

Sewage the waste and wastewater produced by residential and commercial establishments and discharged into sewers.

Sliplining trenchless technology method that slides a new, smaller diameter, polyethelene pipe liner into old damaged pipe.

Sludge the mixture of settleable solids and water removed from the bottom of the settling tank.

Sludge Loading Rate the weight of wet biosolids fed to the reactor per square foot of reactor bed area per hour ($lb/ft^2/h$).

Sludge Retention Time (SRT) See *Mean Cell Residence Time.*

Sludge Volume Index (SVI) a process control calculation that is used to evaluate the settling quality of the activated sludge. Requires the SSV_{30} and mixed liquor suspended solids test results to calculate.

Solids Concentration the weight of solids per unit weight of sludge.

Solids Content (also called percent total solids) the weight of total solids in biosolids per unit total weight of biosolids expressed in percent. Water content plus solids content equals 100%. This includes all chemicals and other solids added to the biosolids.

Solids Loading Rate (drying beds) the weight of solids on a dry weight basis applied annually per square foot of drying bed area.

Solids Recovery (centrifuge) the ratio of cake solids to feed solids for equal sampling times. It can be calculated with suspended solids and flow data, or with only suspended solids data. The centrate solids must be corrected if chemicals are fed to the centrifuge.

Static Head   the actual vertical distance from the system inlet to the highest discharge point.

Steady Inflow   water discharged from cellar and foundation drains, cooling water discharges, and drains from springs and swampy areas. This type of inflow is steady and is identified and measured along with infiltration.

Sterilization   the removal of all living organisms.

Storm Sewer   a collection system designed to carry only storm water runoff.

Stormwater   runoff resulting from rainfall and snowmelt.

Stream Self-Purification   the innate ability of healthy streams (and their biota) to rid themselves of small amounts of pollution. Successful self-purification depends on the volume of water the stream carries, the amount of pollution, and the speed the stream travels.

Supernatant   in a digester, the amber-colored liquid above the sludge.

Surface Water   all water naturally open to the atmosphere, and all springs, wells, or other collectors that are directly influenced by surface water.

Tertiary Treatment   the third step in wastewater treatment, sometimes employed at municipal wastewater treatment plants. It consists of advanced cleaning that removes nutrients and most BOD.

Total Dynamic Head   the total of the static head, friction head, and velocity head.

Total Inflow   the sum of the direct inflow at any point in the system, plus any flow discharged from the system upstream through overflows, pumping station bypasses, etc.

Total Suspended Solids (TSS)   solids present in wastewater.

Transpiration   the process by which plants give off water to the atmosphere.

Trunk lines   collection system pipes that carry wastewater flow to interceptors.

Turbidity   a measure of the cloudiness of water. Caused by the presence of suspended matter, turbidity shelters harmful microorganisms and reduces the effectiveness of disinfecting compounds.

Turbulence   a state of high agitation. In turbulent fluid flow, the velocity of a given particle changes constantly, both in magnitude and direction.

Urban Water Cycle   a local sub-system of the water cycle created by human water use, also called an integrated water cycle. These artificial cycles involve surface water withdrawal, processing, and distribution; wastewater collection, treatment, and disposal back to surface water by dilution and natural purification in a river. The cycle is repeated by communities downstream.

Velocity   the speed of a liquid moving through a pipe, channel, or tank. May be expressed in feet per second.

Velocity Head   the energy needed to keep the liquid moving at a given velocity, expressed in terms of the added system head required.

Vents   collection system ventilation points that ensure that gases that build up within sewer systems from the wastes they carry are safely removed from the system.

Waste Activated Sludge Solids (WASS)   the concentration of suspended solids in sludge removed from the activated sludge process.

Wastewater   the spent or used water from individual homes, communities, farms, or industries that contains dissolved or suspended matter.

Wastewater Collection System   community sewerage system to collect and transport wastewater (1) from residential, commercial, and industrial customers, and (2) from stormwater runoff through storm sewers. Wastewater is transported through the sanitary sewer or combination system to a treatment facility. Stormwater is transported through a storm sewer system or combined system to a treatment facility or approved discharge point.

Wastewater Reclamation   the treatment or processing of wastewater to make it reusable.

Wastewater Reuse   the use of treated wastewater for a beneficial use such as industrial cooling.

Water Cycle   see also *Hydrogeologic Cycle.*

Water Softening   a chemical treatment method that uses either chemicals to precipitate or a zeolite to remove metal ions (typically Ca^{2+}, Mg^{2+}, Fe^{3+}) responsible for hard water from drinking water supplies. The waste by-product is lime sludge.

Waterborne Disease   illness caused by pathogenic organisms in water.

Watershed   the land area that drains into a river, river system, or other body of water.

Weir   a device used to measure wastewater flow.

Wellhead Protection   the protection of the surface and subsurface areas surrounding a water well or wellfield supplying a public water system that may be contaminated through human activity.

Zone of Recent Pollution   the pollution discharge point, where the stream becomes turbid.

Zoogleal Slime   the biological slime that forms on fixed film treatment devices. It contains a wide variety of organisms essential to the treatment process.

Index